CAMBRIDGE EARTH SCIENCE SERI

Editors:
A.H. Cook, W.B. Harland, N.F. Hughes,
J.G. Sclater

Fossil invertebrates

In this series:

Fossil invertebrates

U. LEHMANN *and* **G. HILLMER**

Institute of Geology and Palaeontology
University of Hamburg

Translated by

JANINE LETTAU

University of Cambridge

CAMBRIDGE UNIVERSITY PRESS

Cambridge
London   New York   New Rochelle
Melbourne   Sydney

Published by the Press Syndicate of the University of Cambridge
The Pitt Building, Trumpington Street, Cambridge CB2 1RP
32 East 57th Street, New York, NY 10022, USA
296 Beaconsfield Parade, Middle Park, Melbourne 3206, Australia

Originally published in German as *Wirbellose Tiere der Vorzeit* by Ferdinand Enke
Verlag, Stuttgart, in 1980, and © Ferdinand Enke Verlag 1980

First published in English by Cambridge University Press 1983 as *Fossil
Invertebrates*
© Cambridge University Press 1983

Printed in Great Britain at the University Press, Cambridge

Library of Congress catalogue card number: 82-9419

British Library Cataloguing in Publication Data

Lehmann, U.
Fossil invertebrates. – (Cambridge earth science)
1. Invertebrates, Fossil
I. Title   II. Hillmer, G.
III. Wirbellose Tiere der Vorzeit. *English*
562   QE770

ISBN 0 521 24856 6 hard covers
ISBN 0 521 27028 6 paperback

DJ

Contents

Preface

The word palaeontology (derived from the Greek) signifies the study of ancient life. De Blainville, who used this term in 1825, restricted the interpretation of 'ancient' to the study of fossil organisms. Since that time the term palaeontology has gradually replaced the synonymous term oryctology. Thus generally speaking, palaeontology deals with fossils (from the Latin *fodere*, to dig), remains of organisms which, by certain processes, have become part of the rocks they are embedded in. The analysis of fossils thus presupposes a knowledge of rock-forming processes. In this respect palaeontology combines the biological sciences (zoology and botany) with the 'exact' sciences (chemistry and physics) via geology. Palaeontology shares with geology the historical component in particular.

General palaeontology deals with problems of fossilisation and the relations between the fossil and the rocks containing it. Corresponding present-day processes are dealt with by actuopalaeontology. Special or systematic palaeontology describes the fossils and classifies them into systems. Accordingly, it is subdivided into palaeobotany and palaeozoology; the latter is broken down into the palaeontology of invertebrates and of vertebrates. Micropalaeontology unites palaeobotany and palaeozoology in the study of microscopic organisms. Applied palaeontology, the study of index fossils, has the task of dating rocks with the aid of fossils (biostratigraphy). It is thus an auxiliary science to historical geology. General and special palaeontology provide the basis for studies in the fields of palaeobiology (palaeoecology, palaeoichnology, palaeopathology, etc.).

Although a number of textbooks have been published on general palaeontology in recent years, there is a distinct lack of concise textbooks covering systematic palaeozoology. In this situation, it was initially useful to distribute short notes for teaching, but with increasing numbers attending the courses, it was no longer possible to give each student individual attention. This in turn necessitated more detailed notes and finally led to the writing of this book as the most sensible solution.

This book does not aim to impart the latest knowledge or theories, but rather to provide the learner with a sound introduction to the subject and a guide to further study. It assumes a certain basic knowledge of general palaeontology, geology and zoology. The extensive material is deliberately presented in a simplified and concise form.

We are indebted to a number of colleagues who provided us with valuable information and comments; in particular we would like to thank Dr J.

Kazmierćzak (Warsaw), Professor Dr P. Kaiser and Professor Dr O. Kraus (Hamburg), Dr F. Plumhoff (Wietze), Professor Dr H. Ristedt (Bonn) and Dr M.-G. Schulz (Kiel), as well as Professor Dr G. Alberti, Professor Dr Chr. Spaeth, Professor Dr E. Voigt and Dr W. Weitschat from Hamburg.

Most of the illustrations have been taken from other works, but for the sake of uniformity most have been redrawn by Mrs Doris Lewandowski, whom we would like to thank. The illustrations are designed to convey only a stylised concept of the body plans as an aid to recognition and learning when dealing with collected material. They cannot act as substitutes for collected specimens. The text, illustrations and material, together with directions from the teacher, are designed to draw the attention of the student to pertinent features. In fact, the aim of the book is summarised by these words of J. W. Goethe:

> What is the most difficult of all?
> What seems simplest to you:
> To see with your eyes
> The things before your eyes.

Hamburg, Spring 1980   U. L.
                                                             G. H.

Note added in proof:
For technical reasons the terms planktonic, benthonic and nektonic have not been changed to planktic, benthic and nektic as would have been preferred by the authors.

Introduction

The origin of life

Only a few decades ago the almost total absence of any truly Precambrian fossils and the sudden appearance of large numbers of highly evolved
and diverse animal types at the beginning of the Cambrian was regarded as one
of the most remarkable and puzzling aspects of the Earth's history. The rapid
emergence of new discoveries and new knowledge since then has enabled us to
make more definite statements about Precambrian fossils.

In essence, life could have originated in any of the following three ways:

1. Life could have been created by one or several acts of God, as described
   in the Bible.
2. Life could have been transmitted from the universe in the form of
   bacteria or spores which were transported to Earth by radiation pressure
   or on meteorites and then continued to evolve (cosmozoan or panspermy
   hypothesis).
3. Life could have been created by 'spontaneous generation', i.e. the
   spontaneous (accidental or inevitable) creation of living beings from
   inorganic materials (autogeny) or organic building blocks (plasmogeny).

The first possibility cannot be discussed in terms of the essentially materialistic
methodology of the natural sciences, the second cannot be verified at the present
time, but may well gain importance in the future, while the third is well supported by experiments and fossil finds.

At present, the oldest known fossils are recorded from Greenland: yeast-like
microfossils, about 3800 million years old. Very old fossils are found in the clay
slates and quartzites of the Onverwacht series in the Swaziland system of South
Africa. Their age, determined radiometrically, is about 3400 million years. The
fossils themselves are spherical or thread-like envelopes measuring in the order of
a thousandth of a millimetre. The walls of these envelopes showed traces of
organogenic carbohydrates. In the slightly younger (3000 million years), petro-
graphically similar rocks of the Fig Tree series, fossils have been found which are
thought to be blue-green algae and bacteria (Fig. 1).

The interpretation of the slightly more complex fossils from the bedded and
silicified black slates of the Gunflint series (Ontario, Canada, about 1900 million
years old) is more certain, since these fossils are reminiscent of recent groups of
blue-green algae and even of unicellular and multicellular eukaryotes. Structures
resembling cell nuclei were first identified with some degree of certainty in plant

Fig. 1. Occurrence of Precambrian fossils.

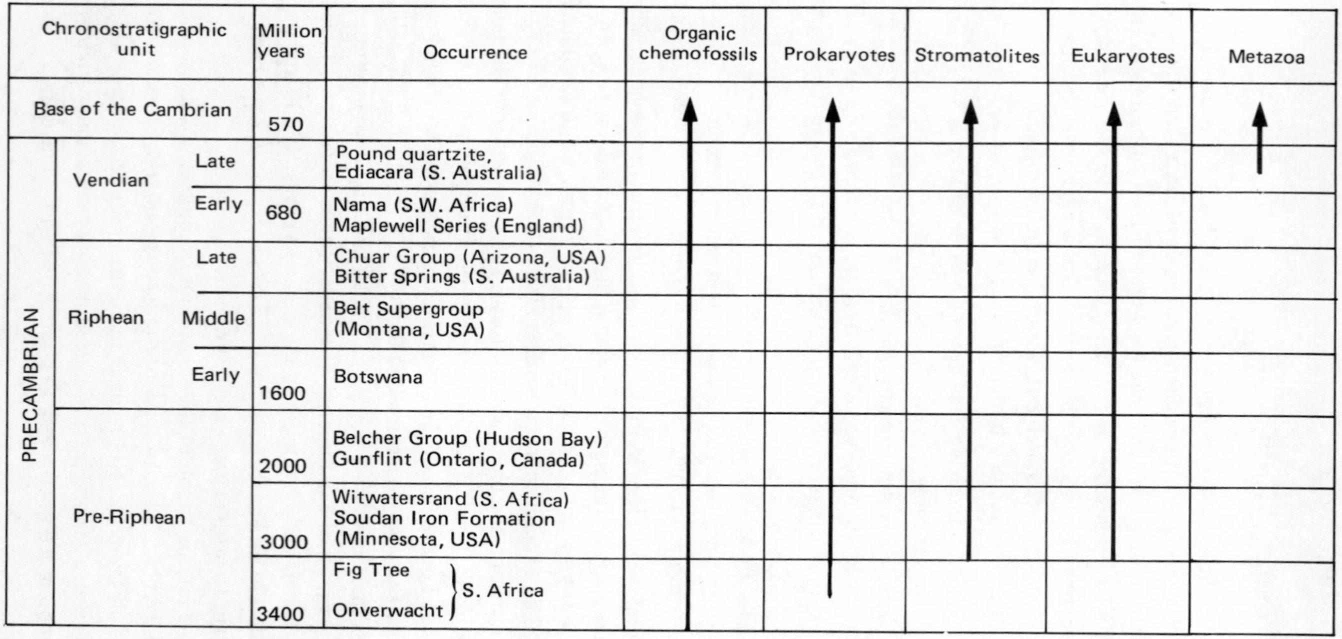

Chronostratigraphic unit		Million years	Occurrence	Organic chemofossils	Prokaryotes	Stromatolites	Eukaryotes	Metazoa
Base of the Cambrian		570		←	←	←	←	←
Vendian	Late		Pound quartzite, Ediacara (S. Australia)					
	Early	680	Nama (S.W. Africa) Maplewell Series (England)					
Riphean	Late		Chuar Group (Arizona, USA) Bitter Springs (S. Australia)					
	Middle		Belt Supergroup (Montana, USA)					
	Early	1600	Botswana					
Pre-Riphean		2000	Belcher Group (Hudson Bay) Gunflint (Ontario, Canada)					
		3000	Witwatersrand (S. Africa) Soudan Iron Formation (Minnesota, USA)					
		3400	Fig Tree } S. Africa Onverwacht }					

PRECAMBRIAN

remains from the rich microflora of the Bitter Springs series in Central Australia (approximately 1000 million years old). There were also signs of mitotic cell division in cells measuring about $10\,\mu m$. The number of finds has continued to increase ever since the technique for finding these oldest of microorganisms was mastered. The technique involves the examination of highly magnified thin sections or polished surfaces, using a fluorescence microscope, for example.

Stromatolites, stratified algal sediments, have been known for some time and were used for stratigraphic purposes because of their time-specific character. The oldest known stromatolites are found in the 2700-million-year-old dolomites of southern Zimbabwe.

The animal and possibly plant fossils known as the Ediacara fauna after their place of discovery in the younger Precambrian of Southern Australia were probably distributed all over the world, given that similar fossils have since been found in a number of other places, notably in South Africa (Nama fossils). Some of the fossils were already tens of centimetres in size and consisted of many cells, but as yet had no supporting skeleton. Some are reminiscent of medusae, octo-corals, echinoderms and worms; others were strange creatures with feathery-striped surfaces. Some would have been autotrophs, others heterotrophs, and some probably had mixed feeding habits (both autotrophic and heterotrophic).

Even though life on Earth is very old, the sudden appearance of most animal phyla in the Cambrian Period is rather surprising. In the Lower Cambrian alone there is a wealth of well preserved protozoans, poriferans, archaeocyathids, coelenterates, brachiopods, annelids, molluscs and various arthropods. Nearly all of them had hard parts. Remains of soft parts have only been found at a few sites with particularly favourable conditions of preservation, e.g. at the Burgess Pass (Middle Cambrian) in British Columbia (Canada).

There is still no generally accepted explanation for this sudden appearance of highly organised, well defined animal types furnished with skeletons. Indeed their capacity and requirement to build preservable hard parts pose questions of their own which have still to be answered.

Since the appearance of Louis Pasteur's award-winning work for the French Academy in 1862, it has been clear that there can be no spontaneous generation in the oxygen-rich atmosphere of the present time, since organic substances are oxygen-depleted and would thus oxidise immediately. Spontaneous generation is only possible in a reducing environment.

According to the geochemists, the primeval hot Earth had no atmosphere. As it gradually cooled off, a primeval atmosphere developed in the course of the removal of gases which resembled present-day volcanic exhalations. It was com-posed mainly of water vapour, carbon dioxide and sulphur dioxide; carbon monoxide, hydrogen, nitrogen, methane and ammonia and traces of other compounds were also present in small quantities, but there was no free oxygen. The atmosphere then was probably comparable with the present-day atmospheres of the giant planets. Oxygen can be liberated from water (or water vapour) and carbon monoxide with a high input of energy which can be supplied by the Sun. There are two possible processes for liberating oxygen: (1) the dissociation of water vapour by ultraviolet radiation in the upper atmosphere, which would yield an oxygen concentration of only about $1^0/_{00}$ of the present-day level; (2) the liberation of elementary oxygen by biological processes (Fig. 2).

Fig. 2. A model of chemical and biological evolution in the Precambrian.

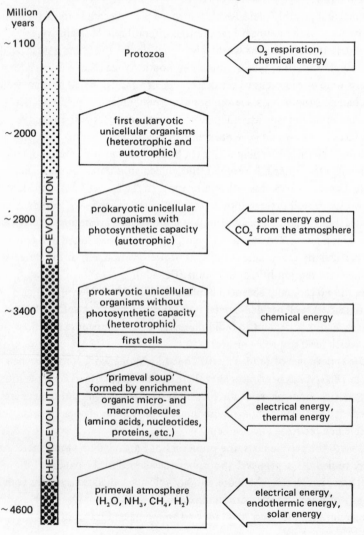

In 1924 A. J. Oparin first showed that an input of energy (lightning, radiation) could bring about the creation of organic substances in a primeval reducing atmosphere: e.g. amino acids (the building blocks for proteins), sugars, purine and pyrimidine bases. A number of other workers were later able to confirm this.

These organic substances were concentrated in water into a so-called primeval soup which became the vehicle for chemical evolution. Compounds of high molecular weight combined to form macromolecules, mainly proteins, nucleic acids and finally polysaccharides. How these substances gave rise to the most primitive life forms whose basic characteristics are metabolism, multiplication (identical reduplication) and the capacity to evolve, has still not been established experimentally. The most important point in this process is the capacity of proteins to produce enzymes and the capacity of nucleic acids to pass on information (Fig. 2).

The most primitive organisms were prokaryotes (without a cell nucleus) which lived anaerobically. They derived their energy by converting glucose to lactic acid, a process which liberates very little energy. In order to escape ultra-violet radiation they must have lived in water at least 10 m deep. In the upper levels of water ultraviolet radiation was of course necessary for the production of amino acids which served as nutrients for the organisms living in the primeval soup. They were thus heterotrophs.

Very early on, organisms evolved which were able to assimilate carbon dioxide and convert it into carbohydrates in the presence of water. Sunlight provided the necessary energy and the newly evolved chlorophyll made the conversion possible. These organisms were thus autotrophs. The process, known as photo-synthesis, yielded free oxygen. Traces of chlorophyll have been found in the Swaziland system (South Africa, more than 3000 million years old). The first and most important producers of oxygen will have been Cyanophyta (blue-green algae).

The oxygen brought about various oxidative processes at the Earth's surface (the oldest red weathering caused by iron oxidation is 1800 million years old) and began to concentrate in the atmosphere. Organisms evolved which were able to respire oxygen and derive their energy from oxygen (the yield of energy in oxygen respiration is 14 times higher than in lactic acid fermentation). At the same time the ozone layer began to form in the upper layers of the atmosphere, which screened the Earth from the ultraviolet radiation of the Sun.

Even today there are some organisms which are able to live either aerobically or anaerobically, depending on whether or not oxygen is present. The transition from one lifestyle to another, the so-called Pasteur effect, occurs when the oxygen content is only 1% of that of the present-day atmosphere (see Fig. 3).

Fig. 3. Surmised increase in the oxygen content of the atmosphere in the course of the Earth's history.

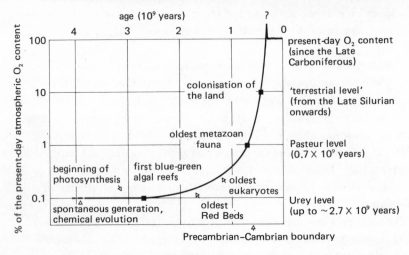

When the critical point (Pasteur point) was reached, there was probably a general transition to oxygen respiration. The resulting increase in available energy brought about decisive evolutionary progress such as the evolution of eukaryotic cells, the manufacture of skeletal materials and more complicated tissue assemblies. The change in fauna in the Lower Cambrian may well be connected with the attainment of a certain oxygen content in the atmosphere.

Eukaryotic cells are considerably more differentiated than prokaryotic ones. In addition to a nucleus, they can possess mitochondria, chloroplasts, specially differentiated flagella and other organelles. Eukaryotic cells may have been created by endobiosis (union) of various prokaryotic cells in a proto-eukaryotic host cell, although other authors postulate differentiation within the cell.

According to current estimates, the oxygen content of the atmosphere in the Upper Silurian, when the first terrestrial plants appeared, was about 10% of the present oxygen level, which was sufficient to screen the surface of the Earth from ultraviolet radiation. Some authors postulate a gradual increase in the oxygen content up to the present day, while others suggest that it reached present levels in the Carboniferous and was then subject to considerable oscillation. Both views are largely speculative.

Bibliography

The following works, which also include extensive lists of references, are recommended for further reading:

Cloud, P. E., Jr (1976*a*) Beginnings of biospheric evolution and their bio-geochemical consequences. *Paleobiology*, 2(4), 351–87.
 (1976*b*) Major features of crustal evolution. *Geol. Soc. S. Afr. Trans.*, Annexure to vol. 79, 32pp.
Degens, E. T. (1977) Physikalische und chemische Steuerungsmechanismen der Evolution. *Ges. Strahlen- u. Umweltforschung, Munich*, 340, 10–46.
Erben, H. K. (1975) *Die Entwicklung der Lebewesen. Spielregeln der Evolution.* R. Piper, Munich & Zurich.
Fairbridge, Rh. W. & Jablonski, D. (1979) *The Encyclopedia of Paleontology.* Dowden, Hutchinson & Ross, Stroudsberg, Pennsylvania.
Flügel, E. (ed.) (1977) *Fossil Algae. Recent Results and Developments.* Springer, Berlin, Heidelberg & New York.
Glaessner, M. F. (1971) Die Entwicklung des Lebens im Präkambrium und seine geologische Bedeutung. *Geol. Rundsch.*, 60(4), 1323–39.
Hahn, G. & Pflug, H. D. (1980) Ein neuer Medusen-Fund aus dem Jung-Präkambrium vom Zentral-Iran. *Senckenbergiana Lethaea*, 60, 449–61.
Havlíček, V. & Kříž, J. (1978) Middle Cambrian *Lamellodonta simplex* Vogel: 'bivalve' turned brachiopod *Trematobolus simplex* (Vogel). *J. Paleontol.*, 52(5), 972–5.
Kaplan, R. W. (1972) *Der Ursprung des Lebens.* dtv-Thieme, Munich.
Kaźmierczak, J. (1979) The eukaryotic nature of *Eosphaera*-like ferriferous structures from the Precambrian Gunflint Iron Formation, Canada: a comparative study. *Precambrian Res.*, 9, 1–22.
Knoll, A. H. & Barghoorn, E. S. (1977) Archean microfossils showing cell division from the Swaziland System of South Africa. *Science*, 198, 396–8.
Margulis, L. (1970) *Origin of Eukaryotic Cells.* Yale University Press, New Haven.
Pflug, H. D. (1971) Neue Fossilfunde im Jung-Präkambrium und ihre Aussagen zur Entstehung der höheren Tiere. *Geol. Rundsch.*, 60(4), 1340–50.
 (1974) Feinstruktur und Ontogenie der jung-präkambrischen Petalo-Organismen. *Paläontol. Z.*, 48(1/2), 77–109.
 (1978) Yeast-like microfossils detected in oldest sediments of the Earth. *Naturwissenschaften*, 65, 611–15.
 (1979) Combined structural and chemical analysis of 3800-Myr-old micro-fossils. *Nature*, 280, 483–6.
Schidlowski, M. (1971) Probleme der atmosphärischen Evolution im Präkambrium. *Geol. Rundsch.*, 60(4), 1351–84.
 (1976) Archean atmosphere and evolution of the terrestrial oxygen budget. *The Early History of the Earth*, ed. B. F. Windley, pp. 525–35. Wiley, New York.
Schopf, J. W. & Oehler, D. Z. (1976) How old are the eukaryotes? *Science*, 193, 47–9.
Sokolow, B. S. (1976) Precambrian Metazoa and the Vendian–Cambrian boundary. *Paleontol. J.*, 10, 1–13.
Vogel, K. & Gutmann, W. F. (1981) *Zur Entstehung von Metazoen-Skeletten an der Wende vom Präkambrium zum Kambrium.* Festschrift d. wissensch. Ges. Joh. Wolfg. Goethe-Univers. Frankfurt a. Main, pp. 517–37, Wiesbaden.

Systematics and the mineralogy of the skeleton

Virtually all invertebrate phyla have been subject to special scrutiny which for various reasons has resulted in the distinction and establishment of a large number of taxonomic entities, particularly genera. What was regarded as a genus half a century ago is now often divided into many genera and can only be viewed as a whole in the framework of one or more families. Sometimes the partition even goes beyond that. As long as it is based on the recognition of certain features (homeomorphisms, heteromorphisms, etc.) the need for detailed classification is undeniable, but the enormous number of names required for classification has a deterrent effect on beginners and also poses problems for the specialist.

The most important aim of this book is to reduce the dread of the large number of names and terms by relating them back to certain fundamental types. All the genera described can and should thus be regarded as collective genera, as examples of certain body plans and their possible permutations. Experts might find it hard to accept the use of generic names like *Perisphinctes* in the section on ammonites or *Productus* in the brachiopods, but given the scope of this book the 'liberal' use of generic names is unavoidable, since it preserves a certain overview which would otherwise be lost in details. This book thus provides only a rough guideline for the identification of fossils, i.e. an indication of where one might obtain more specialised information.

In some cases the ultrastructure and mineralogy of the skeletons are important in systematics (Table 1).

The system of organisms

Organisms are traditionally divided into the two kingdoms of the plants (Plantae) and animals (Animalia). The former are autotrophs (capable of taking up basic organic or inorganic substances by photosynthesis or chemosynthesis), the latter are heterotrophs (capable of utilising preformed organic substances).

The group of unicellular organisms collectively known as Protista is intermediate between plants and animals. These organisms are both heterotrophic and autotrophic, and sometimes one individual may combine both properties. They thus represent an overlap between the plant and animal kingdoms. To create some order in the extraordinary diversity of unicellular organisms, their predominantly autotrophic representatives were assigned to the plant kingdom as protophytes or phytoflagellates and placed in various algal phyla. The predominantly heterotrophic unicellular organisms, on the other hand, remained at the base of the animal kingdom as protozoans. However, this division ran into problems with both the heterotrophic fungi, which had previously been treated as plants, and a group of unicellular organisms without nuclei. These unicellular organisms (prokaryotes) are distinct from all other organisms whose cells do have nuclei, i.e. the eukaryotes.

Table 1. *Skeletal mineralogy of some important fossil groups*

	Aragonite	Calcite	Aragonite + calcite	SiO$_2$	Apatite
Coccolithophorids		X			
Corallinaceans	X	X			
Dasycladaceans	X				
Dinoflagellates	X (?)	X			
Diatoms				X	
Foraminifera	X	X			
Radiolarians				X	
Calcisponges	X	X			
Sclerosponges	X	X			
Hydrozoans	X	X (?)			
Octocorals	X	X	X		
Scleractinians	X				
Pterocorals		X			
Annelids		X			
Bryozoans	X	X	X		
Brachiopods		X			X
Polyplacophorans	X				
Gastropods	X		X		
Scaphopods	X				
Lamellibranchs	X	X	X		
Cephalopods	X		X		
Aptychi		X			
Belemnite rostrum		X (?)	X		
Ostracods		X			
Cirripedes		X			
Echinoderms		X			
Conodonts					X

After Lowenstam (1963), Milliman (1974), Flügel (1978) and others.
A cross represents presence of a compound.

Whittaker (1969) thus suggested that all organisms should be divided into two superkingdoms and five kingdoms, and we have correspondingly classified organisms as follows (see also Fig. 4):

I.   Superkingdom: Prokaryota (unicellular organisms without nuclei)
    1. Kingdom: Monera
    1. Subkingdom: Bacteria
    2. Subkingdom: Cyanophyta (= Cyanobacteria, blue-green algae)

Fig. 4. The sequence of events in evolution.

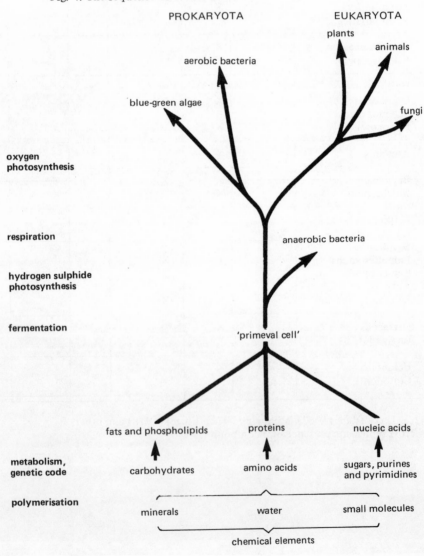

II.　　　Superkingdom: Eukaryota (having cells with nuclei)
　　　　2. Kingdom: Protista (unicellular organisms with nuclei)
　　　　1. Subkingdom: Protophyta (predominantly autotrophic unicellular
　　　　　　organisms, 'lower algae')
　　　　2. Subkingdom: Protozoa (predominantly heterotrophic unicellular
　　　　　　organisms)
　　　　3. Kingdom: Fungi
　　　　4. Kingdom: Plantae
　　　　5. Kingdom: Animalia

This book will deal briefly with the geological significance of the Monera and
Protophyta and in more detail with the geologically important protozoans,
although the main emphasis will be on animals, excluding vertebrates, i.e. on
'invertebrate animals'.

Because of their small size the Monera and Protophyta are mostly part of the
nannofossils (from the Greek *nánnos*, dwarf), while the Protozoa are assigned to
the microfossils. A study of the former requires magnification (usually several
hundredfold) with a microscope, while the latter can be identified with a magni-
fying glass ($\times 5$ to $\times 50$). The terms nannofossil and microfossil have no systematic significance, they are merely based on the practical methods of extraction
and identification. Even among multicellular organisms there are forms which
must be regarded as nannofossils, such as the spores and pollen of plants, or as
microfossils, such as ostracods and conodonts.

I   Superkingdom: Prokaryota
(unicellular organisms without nuclei)

1.   Kingdom: Monera

1.   Subkingdom: Bacteria (Schizomycophyta)

Bacteria are generally less than 1 μm in diameter. They are aerobic or anaerobic and usually heterotrophic or occasionally autotrophic organisms. They are often found in living hosts. Their temperature tolerance is very high, and some bacteria are able to live in hot water up to temperatures of 90 °C. They are killed by direct solar irradiation.

The bacteria are of geological significance because of their important role in the genesis of many ores, notably the Precambrian banded iron ores. They are probably the oldest organisms, and are found in the Fig Tree series, for example (3100 million years; *Eobacterium*).

2.   Subkingdom: Cyanophyta (blue-green algae)

Cyanophytes are diverse, predominantly autotrophic organisms containing chlorophyll and a photosynthetic pigment (phycocyanin). They are usually larger than bacteria, but rarely exceed a diameter of 25 μm. They may unite into branched colonies, so-called trichomes, which are surrounded by a single envelope of cellulose. This envelope is sometimes preserved as a fossil. Small parts of the trichomes may escape from the envelope as hormogonia (singular: hormogonium) and develop independently.

Blue-green algae liberate oxygen in the course of photosynthesis. The pigment phycocyanin reacts to blue light of very low intensity. Cyanophytes are able to exist even when the oxygen content is extremely low and are sometimes anaerobic. They are also resistant to extreme temperatures and drought, but dependent on basic to neutral conditions in their environment (the most acid pH they will tolerate is 4).

Some genera of blue-green algae are responsible for the formation of stromatolites (algal limestones) (Fig. 5), where lawns of algae and thin layers of sediment give rise to very fine bedding. Blue-green algae are also involved in the deposition of travertine. Some promote the chemical and mechanical decomposition of limestones to micrites by virtue of their bore-holes in the limestone.

Fig. 5. Stromatolites from the Precambrian of Southern Australia.

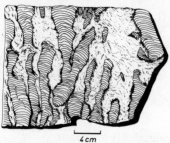

4 cm

The oldest known remains of blue-green algae are described from the Onverwacht and Fig Tree series of South Africa. Stromatolites were particularly diverse in the Upper Precambrian and in the Lower Palaeozoic. Under the appropriate conditions, they are still formed today. These ecologically informative fossils may also serve as guide fossils in Precambrian Strata.

II   Superkingdom: Eukaryota (having cells with nuclei)

Organisms with true cell nuclei.

2.   Kingdom: Protista

Autotrophic and heterotrophic unicellular organisms.

Only those groups of Protista which are of significance as fossils are discussed
here: the phyla Pyrrhophyta and Chrysophyta which are otherwise grouped
with the Thallophyta as 'lower algae', and the phyla Flagellata, Rhizopoda and
Ciliata which are assigned to the animal kingdom as part of the 'Protozoa'.

1.   Subkingdom: Protophyta

Predominantly autotrophic unicellular organisms with cell nuclei.

1.   Phylum: Pyrrhophyta

Pyrrhophytes are predominantly marine, planktonic unicellular organisms
measuring 0.005 to 2 mm. They have chlorophyll and are thus regarded as plants.
They owe their name to the bright red pigments dinoxanthin and peridinin. The
pyrrhophytes, which are (predominantly) autotrophic and heterotrophic, include
the Zooxanthellae which live in symbiosis with corals and large foraminifera.
The Class Dinophyceae is of geological importance on account of its preservable
hard parts.

1.   Class: Dinophyceae (dinoflagellates)

Dinoflagellates occur as a mobile stage or as an encysted stage, and only
the cysts are preserved as fossils. Cysts are known from only a few Recent
dinoflagellates (as resting stages during the winter). Recent dinoflagellates occasionally occur in vast numbers near the ocean surfaces. The wall of the cyst
(phragma) contains cellulose and is chemically very resistant. It consists of two
layers, the outer periphragma and the inner endophragma. The surface is either
smooth or variously ornamented (granular, spinose, or furnished with holes or
a variety of processes). Pylomes, which are thought to be escape holes, are
frequently present (Figs. 6 and 7).

Figs. 6 and 7. Cysts of fossil dinoflagellates. 6: *Gonyaulacysta* (× 400); 7: *Hystrichosphaeridium* (× 400).

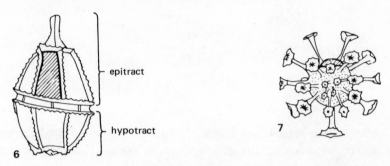

Because of their primitive organisation and the apparent lack of any recognisable relationship with other algal phyla, the Pyrrhophyta are thought to be geologically very old. However, the earliest dinoflagellates described are from the Silurian, and the dinoflagellates as a whole are abundant and of value as guide fossils only from the Triassic onwards. They represent part of the group of nannofossils known as Hystrichosphaeroideae, a heterogeneous group comprising a variety of eggs, spores and cysts. Evitt (1963) gave all the remaining Hystrichosphaeroideae, other than dinoflagellates, the name Acritarcha (p. 37).

2. Phylum: Chrysophyta (yellow-green algae)

The yellowy-green to yellowy-brown colour of the Chrysophyta is attributable to their photosynthetic pigments, to fucoxanthin in particular which masks the green colour of the chlorophyll. Some chrysophytes secrete siliceous or calcareous hard parts and thus become preservable as fossils. The most common and geologically most significant chrysophytes are the silicoflagellates, the diatoms and the coccolithophorids.

1.   Class: Chrysophyceae

1.   Suborder: Silicoflagellata
The silicoflagellates are given the status of a suborder and assigned to the Order Chrysomonadales, Class Chrysophyceae. These unicellular organisms measure 0.02 to 0.1 mm and are partly autotrophic and partly heterotrophic (by way of their extended pseudopodia). Inside, they have a very variable hollow, tube-shaped or rod-shaped skeleton of amorphous silica (opal) (Fig. 8). They inhabit the illuminated surface zones of the oceans down to depths of about 300 m. Fewer than 20 genera are known to have silica skeletons. The oldest silicoflagellates have been found in Lower Cretaceous sediments. Their maximum

Fig. 8. Siliceous skeletons of the silicoflagellates (*c.* × 200).

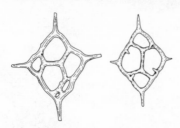

distribution is in the Upper Eocene, the Miocene and Recent, predominantly in diatom-rich sediments. The silicoflagellates are valuable both for biostratigraphic and palaeoclimatic analyses.

Members of the Order Chrysomonadales without skeletons may form cysts impregnated with silica as resting stages, so-called statocysts. Fossil statocysts, known as archaeomonads, also occur in the Cretaceous, while similar structures are already known from Precambrian rocks.

2.   Class: Bacillariophyceae (diatoms, siliceous algae)

Diatoms are unicellular algae which usually measure 0.02 to 0.2 mm, but which may vary between 0.005 and 2 mm in length in exceptional cases. Their tests consist of pectin, a cellulose-like material, and a layer of silica on the outside. The skeleton (frustule) consists of two valves of which the larger (epitheca) overlaps the smaller one (hypotheca) like a lid.

Two orders are distinguished:

1. Centrales (Centricae)

The skeletons are more or less circular to rectangular; the sculpturing of the walls is arranged radially or concentrically. The Centrales are mainly marine planktonic forms (Fig. 9).

2. Pennales (Pennatae)

The skeletons are rod-shaped, boat-shaped or wedge-shaped and often have

Figs. 9 and 10. Skeletons of diatoms. 9: Centrales; 10: Pennales (× 250).

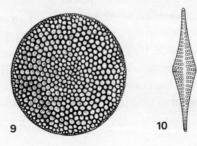

9   10

a non-silicified longitudinal groove (raphe) which is absent in the Centrales. The Pennales are mainly benthic forms inhabiting marine, brackish-water and fresh-water biotopes (Fig. 10).

The centric diatoms are the geologically older diatoms. They predominated during the Cretaceous Period and in the Lower Tertiary, while the pennate diatoms increased dramatically in the Miocene and now dominate. Reports of older (Jurassic or Carboniferous) diatoms are dubious.

In younger sediments diatoms are sometimes rock-forming, giving rise to diatomite such as the siliceous marl in the Pleistocene of North Germany (Kieselgur).

3.   Class: Coccolithophyceae (Coccolithophorida)
        The systematic status of the Coccolithophorida is still uncertain, but it is generally agreed that they should be assigned to the Chrysophyta. They are autotrophic marine planktonic organisms with a mostly spherical gelatinous body covered with calcareous plates (coccolites) (Fig. 11). The diameter of the

Fig. 11. *Coccolithus pelagicus* (diagrammatic representation; *c.* × 500).

plates ranges between 0.002 and 0.01 mm. The plates become detached after death and sink to the sea-bottom where they give rise to thick deposits of calcareous muds or chalk.

Recent coccolithophorids occur in vast numbers in warmer sea areas. Thanks to their diversity, the coccolites are of great biostratigraphic importance. The earliest and then not very numerous coccolites are described from the Triassic. They become important rock-formers from the Jurassic onwards. White chalk, for example, consists mainly of coccolites.

2.   **Subkingdom: Protozoa**
        (predominantly heterotrophic unicellular organisms)

The Protozoa are unicellular, usually microscopic, organisms whose cell body may contain one or more nuclei. They form a number of separate phyla

which are distinguished by their modes of locomotion. Usually only those protozoans with resistant hard parts are preserved as fossils. Because of the large
number of individuals and the high densities thus found at a given site, protozoans have special stratigraphic and ecological significance.

3.   *Phylum: Flagellata* (zooflagellates)

The flagellates are so called because of the flagella they use for locomotion. They may have one flagellum or several flagella arranged in a number of
ways. Even without the predominantly autotrophic (phyto)flagellates (see
pp. 14, 15) the diversity of the remaining predominantly heterotrophic (zoo)-
flagellates is enormous. Since they secrete no preservable hard parts (in contrast
to some phytoflagellates), there are no known fossils, although they probably
did occur in the Precambrian.

4.   *Phylum: Rhizopoda*

The Rhizopoda are unicellular organisms with protoplasmic extensions
(pseudopodia, rhizopodia) which are used for locomotion, feeding and excretion.
They often have more or less rigid tests or skeletons.

1.   Class: Foraminiferida

The foraminifera are predominantly marine unicellular organisms with
tests consisting of one or more chambers. The test may be made up of: (*a*) tectin,
an organic horny substance (formerly occasionally named 'chitin'); (*b*) agglutinated foreign bodies; (*c*) calcium carbonate ($CaCO_3$) in hyaline form (= glassy
translucent), cryptocrystalline form (= like porcelain) or as calcareous granules.
The size of the test may vary between about 0.05 and 150 mm. The protoplasm
concentrated inside the test sends out processes (rhizopodia) through openings
in the wall. In the multichambered (polythalamous) tests the chambers are
separated by septa. The connection between the septa and the wall is usually
recognisable as a suture. The initial chamber is known as the proloculus (plural:
proloculi).

Rolled up tests are called advolute if the whorls touch along a line, evolute if
the whorls touch along a surface, involute if the younger whorls only partly
surround the older ones, and convolute if the older whorls are completely
surrounded.

The following points are particularly important for the taxonomy of foraminifera: the material from which the test is built; position and shape of the aperture;
number, shape and size of the chambers; and the sculpturing of the test (Table 2
and Fig. 12).

Table 2. *Summary of foraminiferan forms and building materials*

Shell shape	Order and building material			
	Textulariida, agglutinated	Fusulinida, granulated-calcareous	Miliolida, calcareous-imperforate	Rotaliida, calcareous-perforate
± spherical	*Saccammina*	*(Umbellina)*	–	*Lagena*
uniserial ± straight	*Reophax*	*(Nodosinella)*	*(Nubecularia)*	*Nodosaria*
planispiral with one chamber	*Ammodiscus*	*(Tournayella)*	*Cyclogyra*	*Spirillina*
planispiral multichambered	*Cyclammina*	*Endothyra*	*Praepeneroplis*	*Elphidium* (→ *Nummulites*)
planispiral uncoiled	*Ammobaculites*	*Haplophragmella*	*Rectocornuspira*	*Dimorphina*
trochospiral	*Trochammina*	*Tetrataxis*	*Barkerina*	*Rotalia* (*Globigerina*)
biserial	*Textularia*	*Palaeotextularia*	–	*Bolivina*
milioline	*Miliammina*	*(Archaediscus)*	*Miliola*	*(Sigmoidella)*
discoidal-fusiform, large	*Orbitolina*	*Fusulina*	*Alveolina*	*Nummulites*

Underlined genera are common, bracketed ones rare.

Fig. 12. Foraminiferan apertures: (1) free end of a spiral tube; (2) basal, oval; (3) simple, round, on the frontal side (areal aperture); (4) terminal, round; (5) crescentic, subterminal; (6) terminal, slit-shaped; (7) tear-drop-shaped, with inner dental plate (bulimin); (8) bottle-shaped, phialine; (9) terminal, with an internal ('entosolenid') tube; (10) dendritic; (11) and (12) sieve-like (cribrate); (13) rhomboid. Not to scale.

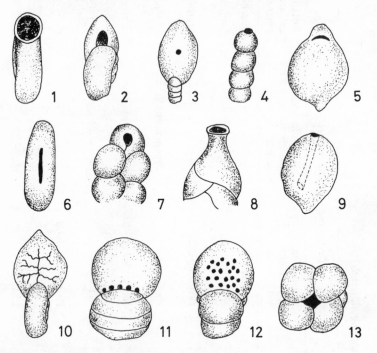

Reproduction occurs by means of a regular alternation between a sexually and an asexually produced generation (generation dimorphism). The latter usually has larger proloculi and is thus known as megalospheric (macrospheric: Fig. 13), in contrast to the microspheric, sexually created generation. However, the size difference is reversed in the adult shell (Fig. 13).

In modern foraminifera the ratio of microspheric to megalospheric individuals of one species is between 1:2 and 1:30 or more. The difference in size between the two generations is particularly marked in the large foraminifera. It is thus possible and usual to divide foraminifera into 'small' and 'large' forms.

Based on the composition of the test, the most important groups are thus:

1. foraminifera with agglutinated tests;
2. foraminifera with tests made up of calcareous granules;
3. foraminifera with calcareous imperforate (cryptocrystalline) tests;
4. foraminifera with calcareous perforate tests ('Hyalina').

Fig. 13. Alternation of generations in the foraminifer *Elphidium*.

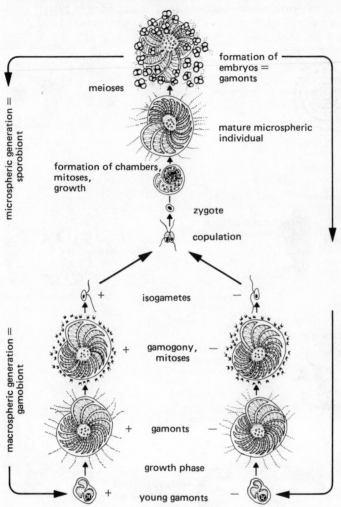

formation of embryos = gamonts

meioses

microspheric generation = sporobiont

mature microspheric individual

formation of chambers, mitoses, growth

zygote

copulation

isogametes

+   −

macrospheric generation = gamobiont

gamogony, mitoses

+ gamonts −

growth phase

+ young gamonts −

1.   Order: *Allogromiida*

Membranous or tectinous test: geologically insignificant. Occurrence: Lower Cambrian–Recent.

2.   Order: *Textulariida*

Elongated and biserial test, either free or sessile; spherical, tube-shaped, coiled up in various ways. Simple wall with a tectinous inner and an agglutinated outer layer. Aperture round, simple and terminal, often absent.

Fig. 14. (a) *Ammodiscus* (×17); (b) *Rhabdammina* (×10); (c) *Reophax* (×18); (d) *Cyclammina* (×4); (e) *Trochammina* (×29); (f) *Loftusia* (top: ×0.7; left: ×22; right: ×3.5); (g) *Cyclolina* (×11.5).

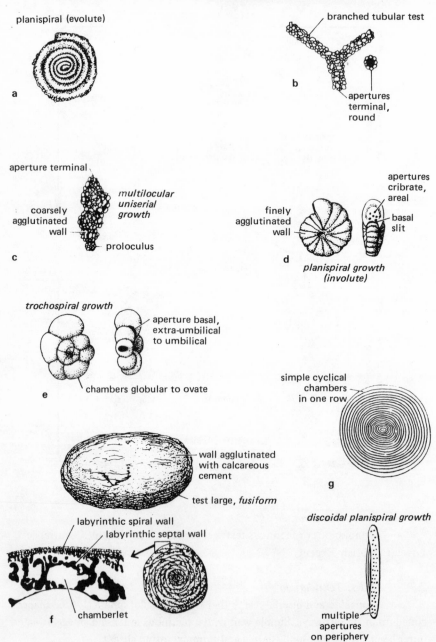

planispiral (evolute)

a

branched tubular test

b

apertures terminal, round

aperture terminal

coarsely agglutinated wall

multilocular uniserial growth

proloculus

c

finely agglutinated wall

apertures cribrate, areal

basal slit

d

planispiral growth (involute)

trochospiral growth

aperture basal, extra-umbilical to umbilical

chambers globular to ovate

e

simple cyclical chambers in one row

wall agglutinated with calcareous cement

test large, *fusiform*

g

labyrinthic spiral wall

labyrinthic septal wall

chamberlet

f

discoidal planispiral growth

multiple apertures on periphery

The Textulariida can be divided into two superfamilies: the Ammodiscacea (mostly unilocular) and the Lituolacea (multilocular).

The Ammodiscacea (Cambrian-Recent) are benthic foraminifera. The genus *Saccammina* (Fig. 19a) (Silurian-Recent) is a simple globular form with a terminal aperture. Tubular tests generally have several apertures radiating out from a central point as in *Rhabdammina* (Fig. 14b) (Ordovician-Recent). Planispiral coiling is seen in *Ammodiscus* (Fig. 14a) (Silurian-Recent).

The Lituolacea are more complex. The simplest benthic forms are straight uniserial, e.g. *Reophax* (Fig. 14c) (Upper Devonian-Recent), or biserial, e.g. *Textularia* (Fig. 19b) (Upper Carboniferous-Recent). Coiled growth patterns are also common, e.g. in the planispiral *Cyclammina* (Fig. 14d) (Cretaceous-Recent) and the trochospiral *Trochammina* (Fig. 14e) (Lower Carboniferous-Recent).

The genus *Loftusia* (Fig. 14f) resembles the more ancient fusulines in having a planispiral fusiform test with a labyrinthine wall, irregular septa and small chambers. *Cyclolina* (Fig. 14g) (Upper Cretaceous) has a discoidal planispiral test with a ring of simple chambers. Conical forms belonging to the Orbitolinidae (Lower Cretaceous-Upper Eocene) have uniserial stacks of saucer-shaped chambers following an early trochospiral stage (e.g. *Orbitolina*, Lower to Upper Cretaceous; Fig. 15). Radial septules subdivide these chambers into an outer zone of small tubular chambers. Within these chambers, smaller horizontal and vertical plates may form a marginal zone of minute cellules. In the centre of the

Fig. 15. Reconstruction of *Orbitolina* (diagrammatic, × 25): 1, marginal zone with cellules; 2, radial zone with chamberlets; 3, reticulate zone.

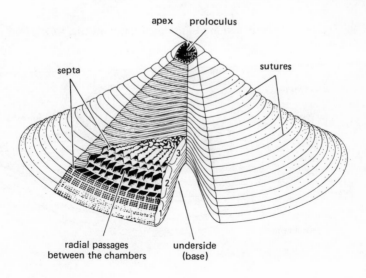

apex   proloculus

septa

sutures

radial passages
between the chambers

underside
(base)

main chamber there is a reticulate zone in which the small radial chambers are further subdivided by vertical pillars.

3.   *Order: Fusulinida*

The extinct fusulines were larger forms with microgranular perforate tests. The chambers were arranged planispirally to a discoidal or fusiform plan. The fusulines have a fusilinid wall structure which is two-layered with mural pores (ancestral type). The schwagerinid wall, on the other hand, lacks a secondary thickening, and the mural pores are enlarged to form alveoli. The schwagerinid wall is typical of the larger fusulines of the Upper Carboniferous and the Permian. Occurrence: Ordovician–Triassic.

The tests are usually of an irregular spindle shape and range between 0.5 mm and several centimetres in size. In addition there are spherical, lens-shaped and subcylindrical forms. The tests are typically planispiral and convolute. The axis of coiling generally coincides with the maximum diameter of the test. Diversity is created by the varying structure of the wall (spirotheca) and the shape of the flat or curved septa. The numerous short chambers are interconnected by pores and channels. The test, which is made up of calcareous granules, is perforated and consists of one to four layers (Fig. 16A).

In simple forms the spirotheca consists of a single undifferentiated layer, the protheca, which may be covered by a thin outer layer, the tectum, and an outer tectorium. In highly differentiated fusulinids (Genus *Fusulina*) the protheca is replaced by the diaphanotheca (a light-coloured translucent layer) under which the dark thin inner tectorium is developed secondarily. The typical four-layered

Fig. 16A. Reconstruction of *Fusulina* (diagrammatic, ×4).

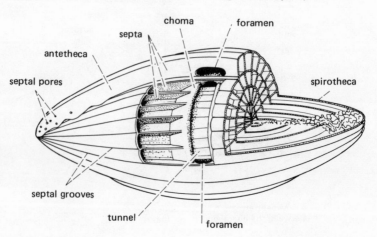

Fig. 16*B*. Foraminifera with fusiform growth. Axial sections.
(a) *Fusulina* (× 7); (b) *Schwagerina* (× 7); (c) *Neoschwagerina* (× 13).

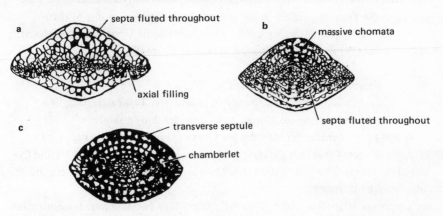

fusilinid wall thus consists of the outer tectorium, the tectum, the diaphanotheca and the inner tectorium.

Fusulina (Fig. 16*B*, a): Test spherical to elongate-fusiform or irregularly subcylindrical; spirotheca generally composed of tectum with upper and lower tectoria, but many genera with diaphanotheca below tectum; septa fluted; chomata massive to weak. Occurrence: Upper Carboniferous.

In the Permian schwagerinids there was a tendency to fill these central, axial chambers with secondary calcite.

In the *Schwagerina* type the wall consists of two layers: the tectum with its fine pores on the inside and the thick keriotheca with its large pores on the outside. The septa are folded in a complicated fashion. In the neoschwagerinids the chambers are subdivided by regular axial and transverse plates, the septules.

Schwagerina (Fig. 16*B*, b): Test fusiform to subcylindrical; spirotheca thick and composed of tectum and alveolar keriotheca; septa fluted; axial infillings highly variable. Occurrence: Permian.

Neoschwagerina (Fig. 16*B*, c): Test large, inflated and fusiform to ellipsoidal; wall thick, composed of tectum and alveolar keriotheca; alveoli to base of septules; one transverse septule to each foramen. Occurrence: Upper Permian.

Ecologically, the fusulinids were purely marine benthic inhabitants of the shallow zones of clear water far from the shore. They are predominantly found in pure limestones, often in association with calcareous algae. The fusulinids first appeared in the Lower Carboniferous and were particularly abundant and wide-spread in the Permian, mainly in the area of the Tethys Sea. Some of them represent very important index fossils: *Fusulina* (Middle Upper Carboniferous), *Schwagerina* (Permian), *Neoschwagerina* (Upper Permian).

4.   Order: Miliolida ('Porcellanea')
        Test coiled either planispirally or irregularly. Test calcareous, like porcelain. Aperture terminal, simple, with small teeth or cribrate.
        Among others this order includes the Genus *Quinqueloculina* (Fig. 19c) (Jurassic-Recent) in which the chambers are added at angles of 144° leaving five chambers visible from the outside (e.g. *Quinqueloculina*). In other genera the angle may be different.
        Larger porcelaneous foraminifera fall mainly into two families: the Soritidae and Alveolinidae. A well known genus of the Family Soritidae is *Orbitolites* (Fig. 20a) (Upper Palaeocene-Eocene), a discoidal planispiral growth form with rings of chambers. The chambers are subdivided into smaller chambers by septules. Occurrence: Carboniferous-Recent.

        Family: Alveolinidae
        The alveolinids have large tests (2-100 mm) which are spherical, ellipsoid or spindle-shaped. Externally, they are similar to the fusulinids of the Lower Palaeozoic. The alveolinid test is imperforate, porcelain-like and bilayered. One can distinguish a primary light-coloured external layer (exoskeleton) and a secondary internal layer (endoskeleton). The internal layer lines the internal surfaces of the chamber walls and makes up the septules (= plates) which are arranged in the direction of coiling. The frontal wall contains numerous openings occurring in one or more rows. *Fasciolites* (= *Alveolina*) (Lower Eocene) is a well known genus (Fig. 17).
        The alveolinids were probably created polyphyletically. They are known to have occurred in the Albian, but are particularly abundant in the Upper Cretaceous and in the Eocene. They are partly rock-forming. Like their present-day relatives, they inhabited shallow tropical waters. Present-day alveolinids live in symbiosis with Zooxanthellae (unicellular green or red algae).

5.   Order: Rotaliida
        Test free, only rarely attached; chambered, originally trochospiral, but other shapes also possible. Walls calcareous and perforated. Test constructed of concentric layers (laminae); each lamina corresponds to one period of chamber construction. Various types of buttress; double septa. Aperture originally slit-shaped, but may be replaced by pores. Occurrence: Permian-Recent.

Fig. 17. Reconstruction of *Fasciolites* (diagrammatic, × 36).

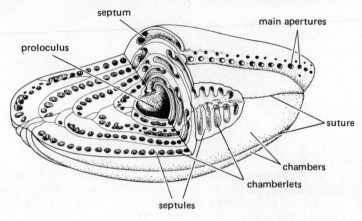

Lagena (Fig. 19d): Unilocular, flask-shaped chamber; costate ornamentation, long neck; aperture terminal. Occurrence: Jurassic–Recent.

Nodosaria (Fig. 19f): Simple uniserial test; terminal radiate aperture. Occurrence: Permian–Recent.

Uvigerina (Fig. 19i): Test elongate, triserial, rounded in section; chambers inflated; wall calcareous, perforate; surface smooth, hispid or costate; aperture terminal, rounded with imperforate neck. Occurrence: Eocene–Recent.

Lenticulina (Fig. 19h): Common involute, planispiral form; lenticular, terminal radiate aperture. Occurrence: Triassic–Recent.

Species of *Globotruncana, Globorotalia, Globigerina* (Palaeocene–Recent) and *Orbulina* (Miocene–Recent) are widely used for correlation.

Globotruncana (Fig. 19k): Trochospiral test; basal umbilical aperture; double keel and beaded sutures.

Globorotalia (Fig. 20e): Trochospiral test; aperture basal, extra-umbilical; single keel and thickened sutures.

Globigerina (Figs. 18 and 19j): Trochospiral test; basal umbilical aperture; inflated chambers.

Orbulina (Fig. 19g): Early trochospiral 'Globigerina' stage; final chamber spherical, embracing earlier ones; aperture of areal pores.

Fig. 18. *Globigerina*, Recent (×70).

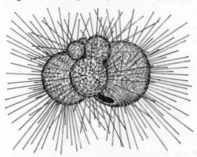

Fig. 19. The skeletons of the following foraminifera: (a) *Saccammina* (×25); (b) *Textularia* (×10); (c) *Quinqueloculina* (×20); (d) *Lagena* (×40); (e) *Elphidium* (×30); (f) *Nodosaria* (×30); (g) *Orbulina* (×15); (h) *Lenticulina* (×15); (i) *Uvigerina* (×40); (j) *Globigerina* (×60); (k) *Globotruncana* (×45).

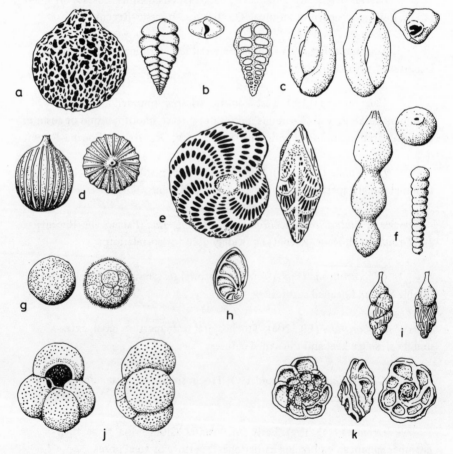

Elphidium (Fig. 19e): Involute, planispiral test; retral processes with canal openings in between; aperture consisting of a row of basal pores. Occurrence: Lower Eocene–Recent.

Ammonia (Fig. 20c): A commonly brackish-water genus; trochospiral test; open umbilicus with pillars and granular sutures. Occurrence: Miocene–Recent.

The nummulites and the orbitoids are large foraminifera with perforated calcareous tests.

Fig. 20. (a) *Orbitolites* (× 7); (b) *Discocyclina* (× 3); (c) *Ammonia* (× 22.5); (d) *Omphalocyclus* (× 10.5); (e) *Globorotalia* (× 15.5).

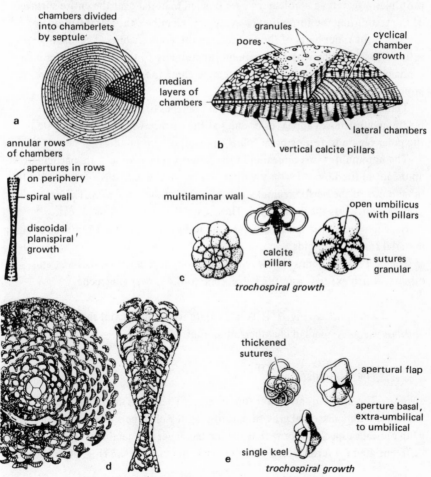

The nummulites

The nummulites are commonly known as 'petrified lenses' or 'coin stones' (from the Latin *nummulus*, small coin). The planispiral tests are lens-shaped or disc-shaped and multichambered with a finely branching system of channels. The tests are calcareous and perforated.

The structure of the nummulites is generally investigated by taking equatorial (horizontal) and axial sections. The latter go through the proloculus at right angles to the equatorial plane. The umbilicus is also known as the polar region. In predominantly involute tests the chambers are either involute or evolute. The lumen of the involute chambers extends towards the pole in elongated processes (alate processes). In evolute chambers the lumen is restricted to the periphery. The external wall of the test is called the spiral lamina (spiral wall). Since the protoplasm secretes a whole new layer of shell material over the entire surface of the test during the formation of every new chamber, the spiral lamina is thicker in the older parts of the test than in the younger ones. The chamber sutures (= septal filaments) correspond functionally to the sutures of the cephalopods. They can be made visible by etching the external wall and are important characteristic features.

Trabeculae are simple or forked lateral branches of the septa. Radially positioned buttresses appear as nodules at the surface which may aggregate in the polar region to form a polar 'plug' ('pustulous' nummulites).

The nummulites first appeared in the Latest Cretaceous and are particularly abundant in the Lower Tertiary with a number of index fossils. They are benthic inhabitants of the neritic regions of tropical and subtropical seas. Their evolutionary centre was the Tethys Sea. These days they live in tropical seas.

Eocene nummulitic limestone was used in Egypt from time to time as building material for the pyramids.

The best-known genus is *Nummulites* (synonyms *Camerina*, *Assilina*, and others), which existed from the Palaeogene to the Lower Oligocene.

Nummulites (Fig. 21): Test involute to evolute, spiral sheet with or without vacuoles; median chambers numerous, simple.

The orbitoids

The orbitoids are larger foraminifera. Their tests are radial hyaline and perforate with a discoidal growth pattern, i.e. the chambers are arranged in rings. Thin sections especially show that the median layer of chambers (equatorial) is differentiated from the lateral chambers, e.g. in *Discocyclina* (Fig. 20b) (Eocene).

Fig. 21. Reconstruction of *Nummulites* (diagrammatic, × 4).

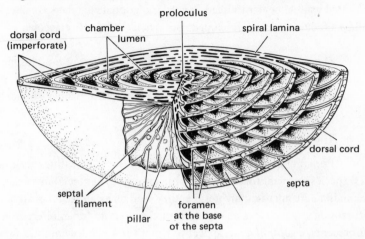

The orbitoids have large disc-shaped or lens-shaped biconcave tests consisting of a single-layered wall. The chambers are arranged in circles in the equatorial plane (equatorial chambers). The equatorial chambers themselves are surrounded by several layers of lateral chambers. In the megalospheric forms the nucleoconch (= embryonal shell) consists of two chambers, the protoconch and the deutero-conch, or several chambers (multilocular). The microspheric generation has a biserial initial stage (the first chambers are not known). The nucleoconch is surrounded by the periembryonal chambers. The equatorial chambers are connected to each other via stolons (tubes). In the layer of lateral chambers, buttresses are often formed which appear as nodules at the surface. When these concentrate in the polar region, they can give rise to a polar 'plug' (so-called pustulous forms), as in the case of the nummulites.

The orbitoids existed from the Campanian to the Maastrichtian; they inhabited the warmer shallow-water regions.

The common genera are *Orbitoides* (Campanian–Maastrichtian) (Fig. 22a, b) and *Omphalocyclus* (Maastrichtian) (Fig. 20d).

Fig. 22. Sections of a specimen of the Genus *Orbitoides:* (a) equatorial section (× 16); (b) embryonic apparatus (× 30).

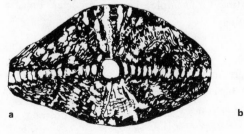

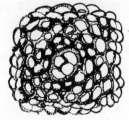

a   b

Orbitoides (Fig. 22): Embryonal chambers surrounded by a thick, perforated wall, bilocular to quadrilocular; equatorial chambers arcuate; lateral chambers reduced or well developed. Occurrence: Upper Cretaceous.

Omphalocyclus (Fig. 20d): Embryonal chambers of megalospheric generation like *Orbitoides*, but with lateral chambers of the same kind and not differentiated from the equatorial chambers. Occurrence: Upper Cretaceous.

The ecology of the foraminifera

Foraminifera are predominantly marine, mobile benthic organisms. A few species are planktonic (and occur in enormous numbers) or sessile. Foraminifera are micro-omnivores, i.e. they feed on small bacteria, algae, protistans or in some cases on organic particles. Certain foraminifera from reef environments, e.g. *Elphidium*, appear to benefit from endosymbiotic algae in the same way as the hermatypic corals do. The fossil 'large foraminifera' probably achieved their great size in this way, with the endosymbiotic algae providing nutrients from photosynthesis and promoting maximum calcium carbonate precipitation by the uptake of carbon dioxide (Brasier, 1980).

Silty and muddy substrates with a rich content of organic debris are favoured by large foraminiferan populations. Certain foraminifera prefer hard substrates to which they temporarily or permanently attach themselves. Some are good facies fossils, being indicators of salinity and water temperature in particular. The cold-water genera generally have agglutinated tests, while the warm-water genera tend to have calcareous tests. The size of the tests may also depend on the water temperature. Thus, foraminifera with agglutinated tests are larger in cold water than they are in warm water; for foraminifera with calcareous tests the reverse is true. In addition, the plankton/benthos ratio permits conclusions about the water depth and temperature. The ratio of planktonic to benthic foraminifera is 9:1 in deep waters, 1:1 at the shelf edge, and less as the coast is approached.

Foraminifera have existed since the Cambrian but became especially abundant and diverse from the Cretaceous onwards. Because of their great biostratigraphic significance (in oil prospecting, for example) they are important objects of research in micropalaeontology.

2.   Class: Actinopoda

The Actinopoda are unicellular organisms with fine, sometimes stiffened, pseudopodia. The majority have a fine skeleton consisting of silica, strontium sulphate or chitinous substance.

1.   Subclass: Radiolaria

Radiolarians are marine unicellular organisms which rarely form colonies. Their protoplasm (= malacoma) is divided into endoplasm and ectoplasm by a central capsular membrane furnished with pores and consisting of tectin.

Most radiolarians have a skeleton (scleracoma) of amorphous silica or strontium sulphate ($SrSO_4$). They vary in size between 0.1 and 0.5 mm, although occasionally some forms may reach 4 mm, and are extremely diverse.

Ernst Haeckel (1834–1919) studied these 'artistic creations of nature' in great detail. In 1860 and 1887 he published monographs on Radiolaria in which he described some 1000 Recent species and illustrated them with over 100 plates which he drew himself.

Radiolarians are distributed in all oceans at all depths. Their diversity is greatest in the Tropics. Cold-water forms are generally larger than warm-water forms. Their lifestyle is planktonic, and they are able to float with the aid of fat bodies and alveoli in the protoplasm as well as by means of the pseudopodia and skeletal projections which increase the surface area.

Radiolarians have been described from Precambrian rocks, but they first became abundant in the Cretaceous. They probably evolved from the dino-flagellates. Their classification is in a state of flux. The Orders Spumellaria and Nassellaria with opaline skeletons are now united in the Superorder Polycystina. Important fossil forms are:

1.   Order: Spumellaria

The silica skeleton is more or less spherical, the capsular membrane perforated by pores (Fig. 23a).

Occurrence: Cambrian–Recent.

Fig. 23. Radiolarian skeletons of the orders (a) Spumellaria (*Actinomma*, × 250) and (b) Nassellaria (*Podocyrtis*, × 100).

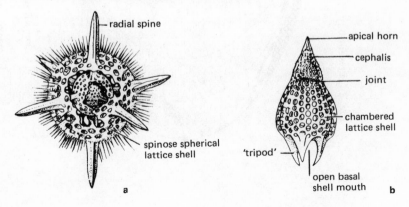

a

b

Fig. 24. Present-day distribution of globigerina and radiolarian ooze.

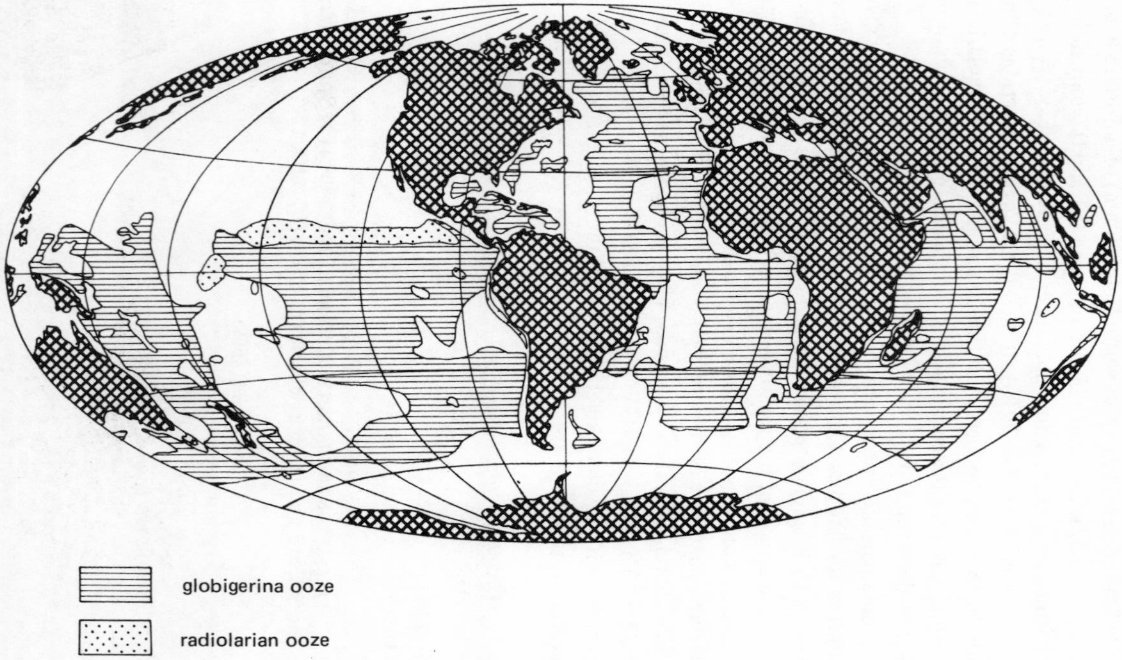

globigerina ooze

radiolarian ooze

Actinomma: Three concentric spinose lattice shells; radial spines unbranched. Occurrence: Devonian–Recent.

2.   Order: Nassellaria
Central capsule perforated only at one pole, single membrane; skeleton in the shape of a tripod, ring or lattice shell (Fig. 23b).
Occurrence: Cambrian–Recent.

Podocyrtis: Conical, segmented skeleton with an apical horn; tripod of three radial spines round the open mouth. Occurrence: Cretaceous–Recent.

2.   Subclass: Acantharia
Although the Acantharia are insignificant as fossils, they are worth mentioning because their skeletal structure diverges from that of other radiolarians, i.e. it is made up of strontium sulphate. They were formerly classified as a suborder of the Order Porulosida. Here they are regarded as a subclass of their own next to the radiolarians in the narrower sense.

In the deep seas of the present time there are thin layers of radiolarian muds at depths of 4000 to 8000 m in the lower latitudes (Fig. 24).
Fossil radiolarites ('flint') consist mainly of radiolarian skeletons and are partly associated with terrigenous sediments. Some radiolarians are thus thought to have evolved in shallow waters. Judging by their geotectonic position and their lithological association, most of the extensively distributed radiolarites must have originated in the deep sea. This is further supported by the types of organisms found in these radiolarites which lack any sign of plants and benthic animals.

5.   **Phylum: Ciliata (Infusoria)**
Ciliates derive their name from the fine hairs (cilia) which partly or totally cover their surface and afford locomotion by all beating in the same direction.
The following are the only significant fossil representatives of the ciliates:

Order: Spirotrichida   Suborder: Tintinnina
The Tintinnina include forms with a solid, usually bell-shaped skeleton, the so-called lorica. From the wide opening at the front end the lorica usually narrows into a pointed caudal process (Fig. 25A). The size of the fossilised skeleton is 0.05–0.2 mm. Modern loricae usually consist of organic material (chitin or xanthoprotein), although some may be agglutinated. Fossil loricae are

Fig. 25*A*. Recent forms of the Tintinnina (left: *Tintinnopsis,* ×400).

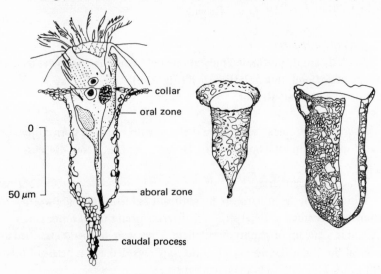

Fig. 25*B*. Longitudinal sections of Tithonian calpionellids (×100).

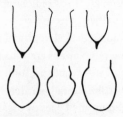

calcareous. According to Remane (in Haq & Boersma, 1978) this rules out any affinity with the ciliates. However, on the basis that the transition from agglutinated to cryptocrystalline calcareous material occurs in other organisms, e.g. in foraminifera, we are leaving them in this systematic position.

Fossil Tintinnina are particularly abundant in the so-called *Calpionella* facies (fine-grain limestones of the Tithonian and the Lower Cretaceous) of the Tethys Sea, where they are also of stratigraphic importance. *Calpionella* is one of the genera from this facies (Fig. 25*B*).

3.   Kingdom: Fungi

Lower plants without chlorophyll; heterotrophs.

H. D. Pflug (1979) described the oldest fungus-like organisms: *Ramsaysphaera* from the Swartkoppie chert of the Swaziland system (3400 million years old) and *Isuasphaera* from the siliceous beds of the Isua series of south-west Greenland (3800 million years old). Both forms are very reminiscent of dehydrated asporogenous yeast cells, even down to their mode of multiplication by budding. However, because of their age (*Isuasphaera* is currently the oldest known organism), Pflug concluded that these fossils are far removed from modern yeasts or other eukaryotes, in spite of their similarity to yeast.

Even though these finds cannot be fully interpreted, there is little doubt that fungi must have developed very early on. However, their role in the fossil record is a subordinate one.

4. Kingdom: Plantae (plants)

Multicellular autotrophic organisms.

Problematical groups (incertae sedis)
1. Group (form group): Acritarcha

The name Acritarcha (from the Greek *ákritos*, uncertain, and *arché*, origin) after Evitt (1963) describes an artificial group of unicellular cyst-like, 0.01–0.5 mm long, nannofossils which sometimes have a multilayered and very resistant wall consisting of an organic substance similar to the sporopollenin of spores. The surface is smooth or sculpted. Pylomes similar to those found in dinoflagellates occur in some Acritarcha. The Acritarcha were formerly grouped together with the dinoflagellates as the Hystrichosphaeroideae. In terms of shape, the Acritarcha are reminiscent of eggs, cysts or pollen of variable systematic origin. According to Brasier (1980) about eight groups of form genera can be distinguished.

Most Acritarcha are part of the marine plankton, although some have also been found in Recent lake sediments. The temperature tolerance of the group as a whole was apparently high. Some assemblages were distributed in parallel with the palaeolatitudes.

Acritarcha first occur in abundance in Upper Precambrian rocks (Riphean). They are very common and valuable as guide fossils in the Lower Palaeozoic, but have become rarer since the Upper Carboniferous.

2. Group: Chitinozoa

The Chitinozoa are club-shaped, bottle-shaped or rod-shaped microfossils with an envelope of a chemically extremely resistant organic material, a diagenetically transformed tectin (pseudochitin). The wide aperture is often furnished with a collar and narrowed by a diaphragm. In some forms the aperture is closed by a lid which extends into a stalk, the so-called copula. The wall consists of the black opaque chamber wall and a thin translucent tegmen.

The systematic status of the chitinozoans is still uncertain. Similarities with the egg cases of worms or gastropods (or even of graptolites?) are thought to indicate an affinity with metazoans, while signs of asexual reproduction (budding) point to a relationship with protozoans or protophytes.

Chitinozoans were purely marine and planktonic or pseudoplanktonic; they occur in a variety of sediments. Finds of Upper Precambrian chitinozoans are dubious. They do not occur in the Cambrian, and their climax was from the Earliest Ordovician to the Devonian. The last representatives are described from the Permian.

3. Group: Petalonamae

Pflug gave the multicellular organisms which measure in the order of centimetres to tens of centimetres, occur in the Upper Precambrian (Vendian) and exhibit traits of metazoans (multicellular animals) with digestive body cavities, the name Petalonamae (from the Greek *petalon*, a leaf, and Namaland in South Africa). In the simplest case, the bodies of the Petalonamae are funnel-shaped colonies of regularly branched tubes which have grown up from the substrate (Fig. 26*A* and *B*). The tubes have fused laterally to form a solid wall

Fig. 26. Petalonamae. *A*. Division and differentiation in *Erniobaris* (juvenile stage): (a) dorsal view, (b) ventral view. Ec, ectodermal wall; En, endodermal wall. (From Pflug, 1974.)

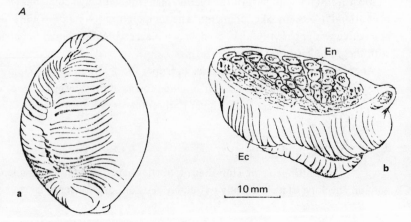

B. Ernietta plateauensis Pflug (1966). Reconstruction (× 0.7) of a group of individuals living on a sandy Precambrian sea-bottom. Juvenile specimens were apparently attached to clumps of mud or other protuberances on the sea-bottom, while more mature specimens may have lived with their basal parts buried in the substrate. (From Jenkins *et al.*, 1981.)

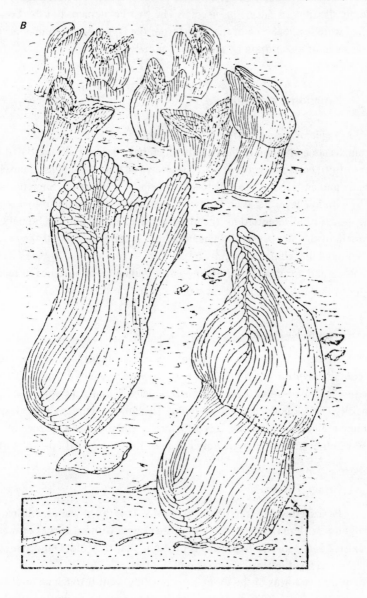

B

which makes up a cup-shaped or urn-shaped envelope opening out at the top. The central cavity (centrarium) which the funnel organisms open into is of diagnostic importance. The external wall has a regular feathery-striped relief pattern. The skeleton consists of an outer organic horny or chitin-like layer and randomly distributed calcareous needles. The first Petalonamae of the Lower Vendian were sessile. Some of the forms from the Upper Vendian may have been capable of locomotion by swimming or crawling.

5. Kingdom: Animalia

Vagile or sessile multicellular organisms.

Animals are composed of many cells which have become differentiated in order to fulfil certain functions. There is always an external cell layer enveloping the body and an internal one which is responsible for digestion. Sometimes there may be a middle layer between the two. Three groups are distinguished according to the degree of differentiation: the Mesozoa, the Parazoa and the Eumetazoa. The origin of the Metazoa is still uncertain, but it seems likely that they evolved from colonial unicellular organisms.

Tables 3 and 4 provide information on the geological importance of most of the groups discussed and on the ratio of fossil and modern species.

1.   Group: Mesozoa

Mesozoa consist of a one-layered external cellular tube and reproductive cells contained within the tube. They may grow up to 7 mm in length. They spend some time at least leading an endoparasitic existence in marine worms, molluscs (especially cephalopods) and echinoderms. This group includes about 50 species, but there are no known fossils. It is also unclear whether the Mesozoa are primitive or whether they have degenerated as a result of their partly parasitic lifestyle.

2.   Group: Parazoa

Parazoa have no true tissues or organs. This taxon was established for the Phylum Porifera (sponges). Whether other phyla without tissues or organs ever existed is not known but it seems probable that they did. Our classification of the Archaeocyathida as Parazoa is presumptive.

The primitive forms of the Eumetazoa possibly went through an undifferentiated parazoan stage, although neither the Porifera nor the Archaeocyathida can be regarded as ascendants (direct ancestors) of the Eumetazoa. The Porifera, on

Table 3. *Geological distribution of various marine invertebrate groups. The height of the curves gives a rough indication of the relative abundance within a given group*

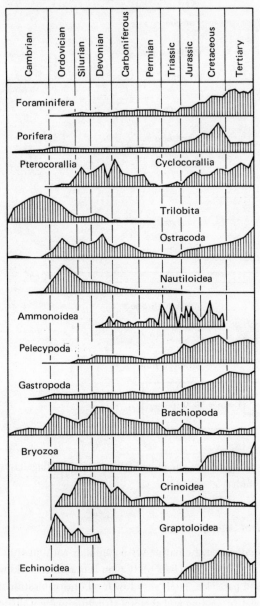

Table 4. *The approximate numerical proportions of Recent and fossil invertebrates*

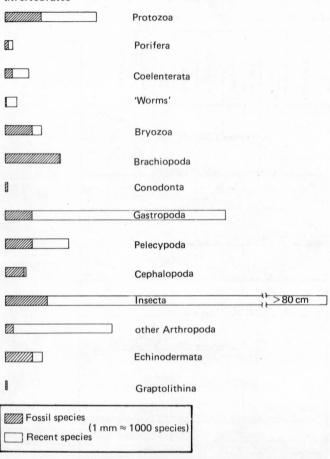

the other hand, can be derived from the choanoflagellates (zooflagellates with a collar round the flagellum).

6.   *Phylum: Porifera* (sponges)

Sponges are predominantly marine sessile animals without true tissues (nervous, muscle or sensory tissues). Their body plan is geared to sucking in and absorbing very small food particles. Their very low level of organisation and their specialisation with respect to feeding and skeletal structure make it difficult to find a direct phylogenetic link between the sponges and any Eumetazoa. In spite of their simplicity they have been successful since the beginning of the Phanerozoic.

Their evolution demonstrates especially clearly that biological success and survival do not depend on the level and degree of organisation or specialisation.

Morphology

The sponge body is composed of three basic cell types:

1. flattened epithelial cells (pinacocytes);
2. collar cells (choanocytes), which line the internal cavities and create a water current by beating their flagella;
3. amoeboid cells which may be differentiated into digestive cells (amoebocytes), skeletal cells (scleroblasts) and reproductive cells.

The amoeboid cells are embedded in a cell-free matrix (colloidal gel). The gelatinous matrix with the embedded skeletal elements, the flattened epithelial cells and the amoeboid cells together form the dermal layer. The internal gastric layer consists of the collar cells. The cells are only in loose contact with one another. They are able to move through the gelatinous matrix in an amoeboid fashion and to transfer nutrients to other cells, particularly to the collar cells (which are apparently able to take up a certain amount of nutrient themselves). It is easy to demonstrate that sponges represent only cell aggregates by putting them through a fine sieve and allowing the fragmented pieces to reaggregate into complete sponges.

The wall of the sponge contains numerous incurrent pores (ostia) through which the water, driven by the choanocytes, flows into a central chamber (paragaster, spongocoel; the commonly used term gastrocoel is misleading, since there is no digestion within this area) and leaves again via a large opening at the top of the sponge (osculum).

One cubic metre of sponge contains about 7500 collar cells. Two hundred grams (dry weight) of sponge drives about 1 tonne of water through its interior in 24 hours.

The lining of collar cells in the paragaster is only sufficient to feed the whole body mass of a small sponge. When an organism increases in size, its surface area increases to the power of 2, while its body mass increases to the power of 3. The number of choanocytes thus has to be increased in the same proportion. This is initially achieved by folding the walls of the paragaster and eventually by creating special chambers with choanocytes. Three grades of organisation represent the most important stages (Fig. 27).

The sponge bodies are attached at their bases. They form new individuals by budding, but these usually remain attached to the main body.

The soft body of the sponges is supported by a skeleton of fibres or needles (scleres, sclerites or spicules). The skeleton consists of spongin (a flexible protein

Fig. 27. Different types of sponge architecture: (a) ascon, (b) sycon, (c) leucon (rhagon) types.

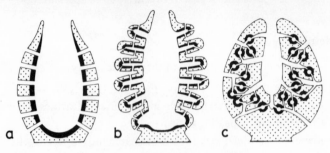

resembling silk), silica (in the amorphous form known as opaline silica) or calcite and/or aragonite (calcium carbonate). The skeletal elements are the only parts of the sponge that may be preserved as fossils and they are thus of great taxonomic importance. Two groups are distinguished, but they are not always clearly defined:

1. Megascleres: relatively large supporting or skeletal spicules, 0.1 to well over 1 mm long, which may fuse to form a coherent framework;
2. Microscleres: small spicules, 0.01–0.1 mm long, which are scattered all over the body. They are hardly ever fossilised, but have taxonomic value for the neozoologist.

Nearly all scleres have an axial canal. They are extremely diverse, both in terms of shape and size. The following basic types of scleres are distinguished (Fig. 28):

1. Monaxons: forms with one axis. If they grow in one direction only, they are known as one-rayed (monactinal); if they grow in both directions they are called two-rayed (diactinal). Styles have one sharp and one blunt end. Tylostyles are pointed at one end and club-shaped at the other end.
2. Tetraxons (or tetracts): four-rayed forms ('tetrapods') in which the four rays emanate equally from the centre. A tetraxon with axes of equal length is also known as a calthrop. Most of the desmas (see below) can be derived from regular tetraxons.
3. Triaxons (= hexacts; three axes; six-rayed): these consist of three axes crossing each other at right angles. The hexact has smooth straight rays which gradually run into points. The pinule has numerous distal-pointing spines or nodes on the branch and thus looks like a coniferous tree. Hexacts with lantern-shaped, perforated centres are known as lychnisks (lantern nodes). Triaxons form regular skeletal lattices and are developed only in species of the Hexactinellida (siliceous or glass sponges).

Fig. 28. Characteristic sponge spicule shapes.

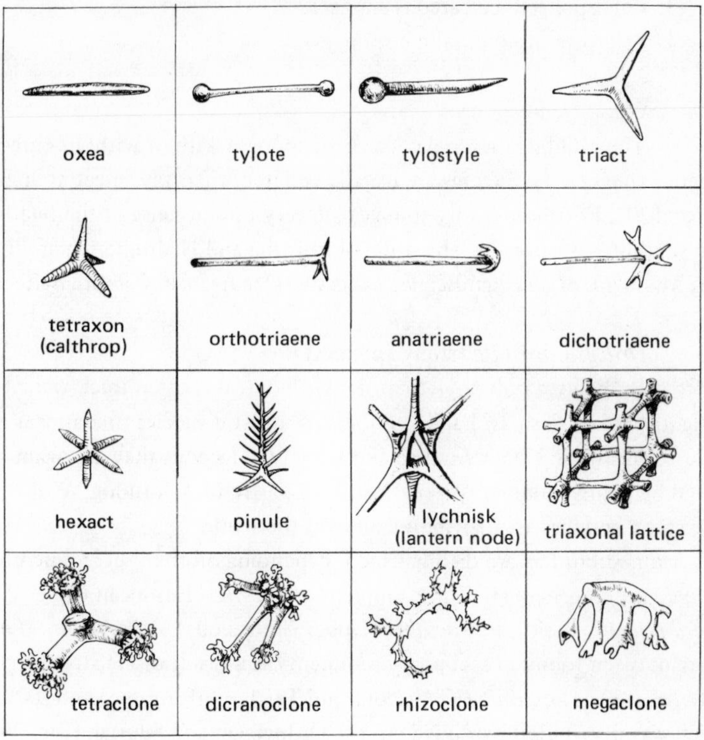

oxea	tylote	tylostyle	triact
tetraxon (calthrop)	orthotriaene	anatriaene	dichotriaene
hexact	pinule	lychnisk (lantern node)	triaxonal lattice
tetraclone	dicranoclone	rhizoclone	megaclone

4. Desmas: skeletal spicules of irregular shape with side branches having root-like ends (= clones). The irregular shape allows them to interlock with neighbouring spicules without fusion. The following types of desma are distinguished: tetraclone, rhizoclone, dicranoclone and megaclone.
5. Polyaxons (multirayed): spherical to star-shaped scleres with many rays; they are usually microscleres.

The classification of the Porifera is largely based on the skeletal material, the spicule type and the skeletal structure. The external shape of the sponge is very dependent on the biotope and not necessarily of prime importance in classification. If the skeletal material is changed pseudomorphically, the shape of the spicules alone determines the classification.

Four classes are distinguished:

1. Demospongea ('common sponges'): horny sponges, sponges without skeletons, and sponges with siliceous spicules other than triaxons.

2. Hyalospongea (= Hexactinellida): siliceous or glass sponges.
3. Calcispongea: calcareous sponges.
4. Sclerospongea: coralline sponges.

1.   Class: Demospongea
        The building material consists of spongin with or without embedded
siliceous spicules. The scleres are usually isolated; their rays meet at angles of
60° or 120°. Fossilised horny sponges are very rare because of the relatively
high solubility of spongin. The orders Lithistida and Hadromerida are the only
ones which are of any significance as fossils. Occurrence: Cambrian–Recent.

1.   *Order: Lithistida* (stony sponges)
        Probably a polyphyletic order with massive, often thick-walled skeleton
of tightly linked desmas. Large numbers of lithistid species first appear in the
Ordovician. In the Upper Jurassic and Upper Cretaceous they are again represented by a large number of genera and are partly rock-forming. At the present
time stony sponges are distributed all over the world.
        Various suborders are distinguished, depending on the type. Some of the
more common genera are, for example: *Callopegma* (European Upper Cretaceous),
Leiodorella (Jurassic), *Verruculina* (Upper Cretaceous), *Aulocopium* (Ordovician-
Silurian; often found in secondary sediments as a result of Pleistocene glacial
drift), as well as *Siphonia* (Cretaceous and Tertiary of Europe) and *Astylospongia*
which was distributed worldwide in the Ordovician and Silurian (Fig. 29).

2.   *Order: Hadromerida*
        Representatives of this order have monaxonal pin-shaped megascleres
(= tylostyles). The order includes the Family Clionidae (boring sponges), representatives of which have been in existence since the Ordovician. Boring sponges
etch labyrinthine passages resembling strings of pearls into inorganic and organic
calcareous substrates, leaving the individual chambers interconnected. Small
holes lead to the surface of the substrate. The Recent genus *Cliona* first
appeared in the Devonian.
        The 'bath sponge' *Euspongia officinalis*, which consists of spongin, belongs
to the Order Keratosida, which is insignificant in terms of fossils.

2.   Class: Hyalospongea (= Hexactinellida)
        This class includes sponges with triaxonal siliceous spicules which meet
at so-called intersection nodes. The rays of the spicules usually form right angles.
Occurrence: Cambrian–Recent.
        Fossil representatives of this class belong to the following orders:

Fig. 29. Sponges of the Order Lithistida: (1) *Aulocopium* (a: × 0.25; b: × 30); (2) *Siphonia* (a: × 0.25; b: × 0.5); (3) *Callopegma* (a: × 0.4; b, c: × 30); (4) *Aulaxinia* (× 0.6); (5) *Astylospongia* (a: × 0.5; b: × 50).

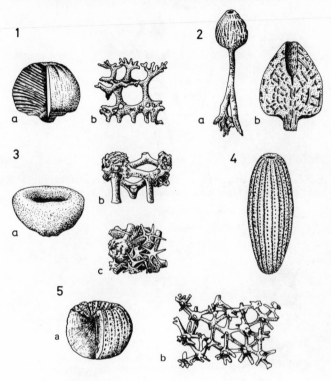

1. Order: Lyssakida

Spherical to cylindrical forms with skeletons consisting mainly of four-rayed (tetractinal, derived from the triaxon by the reduction of one axis) spicules arranged in one plane (so-called stauracts). The skeletal elements are normally not fused, so that the possibilities of preservation are rather limited. However, this order has many Recent representatives. *Protospongia* (Lower Cambrian–Ordovician) and *Hydnoceras* (Silurian–Carboniferous) are representatives of some of the fossil genera (Fig. 30).

R. E. H. Reid (1950) put the Palaeozoic representatives of the Lyssakida into the separate Order Reticulosa.

2. Order: Dictyida

The skeleton consists of a regular lattice of adjacent triaxons. The members of this order often have a well defined funnel shape or vase shape with a wide osculum. They were rare in the Palaeozoic, but numerous and widely

Fig. 30. Sponges of the Order Lyssakida: (1) *Protospongia* (a: × 0.15; b: × 2); (2) *Hydnoceras* (× 0.5).

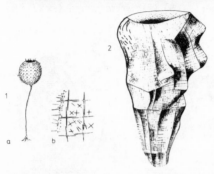

Fig. 31. Sponges of the Order Dictyida: (1) *Tremadictyon* (a: × 0.2; b: × 1; c: × 4); (2) *Guettardiscyphia* (× 0.33).

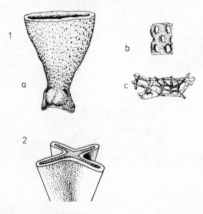

distributed from the Mesozoic onwards. Examples: *Tremadictyon* in the Upper Jurassic and *Guettardiscyphia* from the Upper Cretaceous to the Lower Tertiary (Fig. 31).

3.   Order: Lychniskida

    The Lychniskida, like the Dictyida, also have a regular skeleton, but made up of lychnisks (lantern nodes): triaxons with a lantern-shaped perforation in the centre. In some genera the wall is folded in an undulating fashion. Representatives of this order first appear in the Jurassic. They are particularly abundant in the Cretaceous. Fossil forms are known only from European localities; Recent forms are very rare.

    Ventriculites and *Coeloptychium* from the Upper Cretaceous of Europe are common genera (Fig. 32).

Fig. 32. Sponges of the Order Lychniskida: (1) *Ventriculites* (a: × 0.25; b: × 0.5; c: × 6); (2) *Coeloptychium* (a, b: × 0.33; c: × 30).

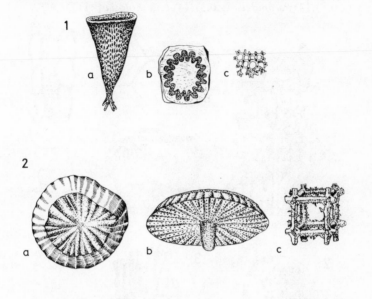

3.   Class: Calcispongea

The skeletal material of these sponges, which have existed since the Cambrian, is exclusively calcareous and consists almost entirely of simple two-, three- or four-rayed spicules. Three-rayed forked scleres are characteristic.

Although the Calcispongea are relatively rare in the Palaeozoic, they appear with large numbers of species in the Alpine Triassic (e.g. St Cassian, Tirol) and in the Cretaceous and Jurassic, but become rarer from the Tertiary onwards.

The following are significant as fossils:

1.   Order: Pharetronida

Calcareous sponges which first appear in the Permian and which have a skeleton of triacts (three-rayed spicules) shaped like tuning forks. Characteristically they have walls of anastomosing fibre tracts and embedded calcareous spicules. For a long time people argued over whether the calcite matrix consisting of fine fibres (sclerosome; pharetrone fibre) existed during the lifetime of the sponge or whether it was created diagenetically after death. Recent 'pharetronid' finds are comparable with fossil pharetronids only with certain reservations. Occurrence: Permian–Recent.

Some of the more common genera are: *Elasmostoma* (abundant in the Cretaceous), *Peronidella* (Triassic–Cretaceous), the spherical *Porosphaera* (Cretaceous of Europe) and *Enaulofungia* (Jurassic–Cretaceous) (Fig. 33).

Fig. 33. Calcareous sponges: (1)–(3) Order Pharetronida; (4) Order Thalamida. (1) *Enaulofungia* (a: ×10; b: ×100; c: ×1); (2) *Elasmostoma* (×0.5); (3) *Peronidella* (×0.5); (4) *Polytholosia* (×1).

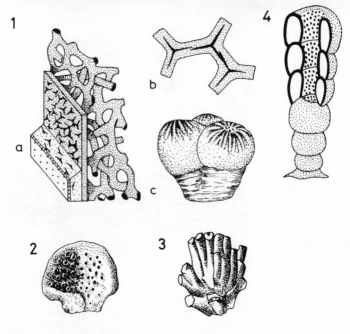

2.   *Order: Thalamida (Sphinctozoa)*

These sponges are built up of more or less spherical chambers arranged in rows. The growth forms are usually reminiscent of strings of pearls. Occurrence: Carboniferous–Cretaceous. Representatives of this order occur mainly in the shallow sea facies of the Permian, Triassic (e.g. the reef-builders of the Alpine Triassic) and the Cretaceous. *Girtyocoelia* (Permian), *Polytholosia* (Triassic) (Fig. 33) and the similar *Barroisia* (Cretaceous) are well known genera.

4.   Class: Sclerospongea

The Sclerospongea, which have existed since the Ordovician, have a calcareous skeleton consisting of tylostyles (pin-shaped spicules) with whorls of thorns and star-shaped branching canals on or directly under the surface (so-called astrorhizae), and many wart-shaped nodules ('mamelons').

The members of this class greatly resemble the stromatopores. In spite of a number of matching characteristics their relationship with the Class Demospongea is still not clear. The Class Sclerospongea was established as a result of studies of Recent forms in the reef areas of Jamaica. It includes the following orders:

1.   Order: *Ceratoporellida*

Solid calcareous skeleton with cup-shaped depressions on the surface and spongious fibrils as well as siliceous spicules. *Ceratoporella* (from Jamaica) is a Recent genus, while *Neuropora* which used to be assigned to the Bryozoa is a fossil representative (Jurassic–Cretaceous).

2.   Order: *Tabulospongida*

The skeleton consists of calcareous tubes with pores. Thorny pseudosepta and numerous tabulae as well as siliceous spicules (monaxons and microscleres) are characteristic. *Acanthochaetetes* is known from the Cretaceous and Recent. Occurrence: Cretaceous–Recent.

3.   Order: *Chaetetida*

The skeleton consists of calcareous tubes without septa, but with siliceous spicules (monaxons) in the tube walls.

The taxonomic status of this order was uncertain until recently. The genus *Chaetetes* existed from the Ordovician to the Carboniferous and was originally regarded as a tabulate coral. The similar genus *Chaetetopsis* (Fig. 34) existed from the Jurassic to Cretaceous periods. Occurrences of the order: Ordovician–Tertiary.

Fig. 34. *Chaetetopsis* of the Order Chaetetida (Class Sclerospongea): (a) complete view (× 0.3); (b) transverse section showing thorns (Th.) (× 6); (c) longitudinal section showing tabulae (T.) (× 6); (d) longitudinal section showing siliceous spicules (monaxonal spicules, m.S.) (× 70).

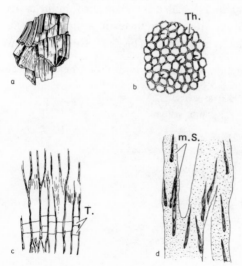

4.   *Order: Muranida*

The skeleton consists of calcareous pillars with siliceous spicules (monaxons) in their axes. The genus *Murania* is known from the Cretaceous.

The ecology of sponges

Fossil sponges often built structures resembling reefs, e.g. in the sponge facies in the Malm (Upper Jurassic) of Southern Germany. They were predominantly inhabitants of neritic regions.

The Recent stony sponges of the Order Lithistida are distributed worldwide below the photic zone at a depth of 100–400 m.

The Recent calcareous sponges (Class Calcispongea) inhabit warm shallow waters with an optimum water depth of 4–18 m. They tolerate neither stagnant nor too agitated waters. Similar requirements can be assumed for the fossil calcareous sponges of the so-called pharetronid facies.

Members of the Recent Class Hyalospongea, the siliceous or glass sponges, are found from the tidal zones down to depths of *c.* 6000 m. They are most common at depths of between 200 and 500 m where they grow on muddy substrates. All other sponges seem to prefer solid substrates.

7.   *Phylum: Archaeocyatha (Archaeocyathida)*

The Archaeocyatha form a diverse group of marine animals restricted to the Lower and Middle Cambrian. Morphologically, they were intermediate between the Coelenterata and the Porifera. Their calcareous skeleton is characteristically cone-shaped, usually double-walled and granulated. In the cavity between the outer and the inner wall, the intervallum, there are vertical and radial pseudosepta (= parieties). The skeleton is perforated (Fig. 35). The soft parts were pre-

Fig. 35. Reconstruction of an archaeocyathid of the Class Regulares (× 10).

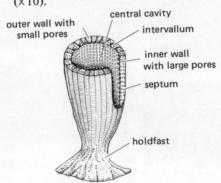

sumably restricted to the intervallum and the surfaces of the walls. There are lamelliform exothecal outgrowths ('holdfasts') at the tip of the inverted cone which presumably served to anchor the animal to the substrate. The diameter of the 'cup' varies between 1.5 and 6 cm, while the height varies between 1.5 and 10 cm.

The Phylum Archaeocyatha is divided into two classes:

1.   Class: Regulares
       Solitary, rarely colonial, cup-shaped to cylindrical forms with a pre-dominantly simple perforated wall which may be more or less spongy in character.

2.   Class: Irregulares
       Predominantly solitary forms with cup-shaped to disc-shaped, often irregular, form. There are usually two calcareous perforated walls which are connected by parieties and irregular porous calcareous plates. Imperforate arched plates (dissepiments) are present; horizontal calcareous plates (tabulae) may be developed later. The forms illustrated in Fig. 36 belong to this class.

Fig. 36. Archaeocyathids of the Class Irregulares: (a) *Tabellaecyathus* (×1); (b) and (c) *Pycnoidocyathus* (× 3).

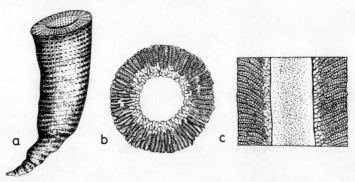

Biological status

The biological status of the Archaeocyatha can only be surmised. Some features are vaguely reminiscent of corals, e.g. the radial parieties, the isolated tabulae and the common cup shape. Other features, on the other hand, such as the large central cavity and the perforated walls, are more sponge-like. However, no skeletal spicules of any sort have been found, and the central cavity is not compatible with the architecture of the corals.

The structure and function of the soft body is also subject to speculation. An evolutionary link between the Archaeocyatha and forms of the Ediacara biocenosis is not particularly convincing either. Their classification as Parazoa seems to be most compatible with the known and reconstructable skeletal structures.

Ecology

The Archaeocyatha were distributed worldwide in a wide warm tropical sea belt. Optimum living conditions were at depths of *c.* 20 to 30 m where they were often associated with blue-green algae. There are detailed studies of Archaeocyathida–Algae bioherms in the Siberian Cambrian. Often there are also trilobites and brachiopods in the immediate vicinity.

A few hundred archaeocyathid species are known mainly from North Africa, Siberia, Australia, North America and the Antarctic. In places they are valuable index fossils.

3.   Group: Eumetazoa

Animals with two (or three) germinal layers in the earliest developmental stages which give rise to certain tissues and organs in the course of ontogeny.

1.   Subgroup: Coelenterata

The main characteristic of the multicellular coelenterates is the single central body cavity (from the Greek *coel enteron*, hollow intestine). Contrary to all other eumetazoans, a secondary body cavity (coelom) is absent. The coelenterates have no central nervous system (with the exception of the nerve ring in the medusae) and no anus. Respiratory, excretory and circulatory systems are all absent.

In the simplest case the 'coelenteron' is a wide sac which communicates with the outside world through an opening acting as both mouth and anus. The body consists of ectoderm and endoderm, which are either in direct contact with each other or separated by a gelatinous supporting lamella, the mesogloea (Fig. 37).

Lateral evaginations of the body cavity are common in medusae, and their function is comparable with that of the blood vessels in higher metazoans, as indicated by the term 'gastrovascular system'.

The Coelenterata are known to have existed since the Precambrian, but only one of the two coelenterate phyla, the Cnidaria, has provided any fossils. The

Fig. 37. Organisation in a scyphomedusa (diagrammatic).

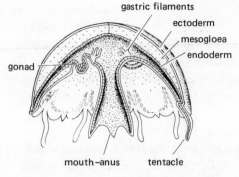

mouth–anus   tentacle

Phylum Acnidaria (Ctenophora, comb-jellies) will therefore not be discussed here. The acnidarians possess colloblasts (lasso cells) instead of nematocysts (stinging cells, cnidae) and move by ciliary action.

8.   Phylum: Cnidaria

The Cnidaria are primitive, predominantly marine coelenterates with radial or bilateral symmetry. The mouth is surrounded by a ring of tentacles. The cnidarians have ectodermal stinging cells (cnidae or nematocysts) which are used for catching prey and for defence. They multiply sexually or by budding. Their lifestyle is solitary or colonial, free-swimming (medusa) or sessile (polyp). During sexual reproduction free-swimming ciliated planula larvae are produced. There is usually an alternation of generations between sessile polyps and free-swimming medusae. Polymorphism within a colony with a division of functions is also common (feeding polyps, reproductive polyps, etc.).

The medusae consist of a water-rich gel. The upper side, the exumbrella, and the underside, the subumbrella, may widen at the margin into a muscular band, the velum (also known as the craspedon). Medusae with a velum are known as craspedote, those without a velum as acraspedote. From the centre of the sub-umbrella hangs the tube-like manubrium with the mouth at its tip. Locomotion is effected by a recoil mechanism. There are longitudinal and transverse muscles. Some medusae have a nerve ring, pigment spots which acts as light-sensitive organs, and statocysts (gravity sense organs).

1.   Class: Scyphozoa

The scyphozoans are marine Cnidaria with tetramerous symmetry. All scyphozoan polyp generations (= scyphopolyps) are very similar, consisting of single polyps 1–7 mm long. The scyphopolyps have four gastral folds (septa).

The medusae have no velum and are thus acraspedote. The periphery is lobed, and the velum is functionally replaced by a subumbrellar muscular mass. The alternation of generations from polyps to medusae occurs by a special type of horizontal budding, so-called strobilation, which in this form is only seen in the Scyphozoa. The medusoid larvae (= ephyra larvae) develop from the polyps, which are divided transversely in several places. In addition to strobilation, the scyphozoans also reproduce by the budding of polyps.

1.   Subclass: Scyphomedusae

The Subclass Scyphomedusae includes both medusae, which can reach a maximum diameter of 2 m, and the very small corresponding polyps, i.e. the Recent Scyphozoa and the few known fossil remains of medusae. In this respect the name Scyphomedusae is misleading. The gastral area (= enteron) is divided into four peripheral gastral pockets by four radial folds pointing inwards. There are lobes along the periphery with interspersed rhopalia (sensory bodies) and tentacles. The mouth is often developed in the shape of a cross. Most species are dioecious. The eggs give rise to ciliated planula larvae.

Since there are no hard parts fossil Scyphomedusae are known only as impressions. The earliest Scyphomedusae are described from the Upper Precambrian, when medusae were diverse and apparently distributed worldwide (Hahn & Pflug, 1980). The lithographic limestone in Solnhofen (Malm) is a classic locality for superbly preserved forms. The best known genus is *Rhizostomites*; virtually its whole anatomy is known in detail. *Aurelia* and *Chrysaora* (compass jellyfish) are examples of Recent genera.

2.   Subclass: Conulata

The Conulata have a conical skeleton, usually 6–10 cm long, of tetramerous symmetry, consisting of a thin chitinophosphatic wall. The outside of the shell is usually covered with fine vertical and horizontal stripes, and each lateral surface is divided into two halves by a more or less marked groove. Although the juvenile stages were predominantly attached to a substrate by their closed apical end, some forms seem to have gone over to a free-swimming lifestyle during the course of further growth. The mouth of the conulariids was very probably surrounded by numerous tentacles and could be closed by more or less triangular lobe-shaped projections of the mouth border.

Occurrence: Cambrian-Triassic. *Conularia* is a genus which was distributed worldwide (Upper Cambrian–Permian) (Fig. 38a and b).

For a long time the conulariids presented a problem to taxonomists. In 1937 they were classified as a scyphozoan polyp generation for the first time, on the basis of their tetrameral symmetry. More recently there have been similar

Fig. 38. (a) and (b) Life-like reconstructions of the genus *Conularia* (× 0.5); (c) Recent genus *Stephanoscyphus* (× 1).

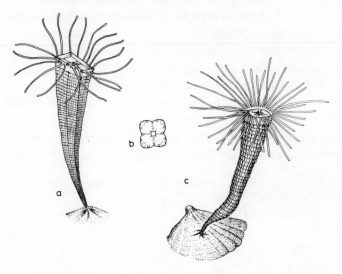

indications in the Recent scyphozoan genus *Stephanoscyphus*, whose polyps are completely surrounded by a chitinous peridermal tube (Fig. 38c).

2.   Class: Hydrozoa

Hydrozoans are predominantly marine Cnidaria with an undivided gastral cavity. The mouth is surrounded by tentacles, but the mouth disc has no gullet (stomadaeum, as in the Anthozoa). The generations generally alternate between the mostly colonial polyp generation and the medusa generation. With a diameter of 2-6 mm the medusae are relatively small; they are craspedote.

Recent hydrozoans inhabit depths of up to 8000 m. The hydrozoan orders of the Milleporida and Stylasterida are of palaeontological significance.

1.   *Order: Hydroida*

The ectoderm of the hydroid polyps usually secretes a chitinous, occasionally a calcareous, exoskeleton (= periderm). In colonial hydropolyps the exoskeleton often covers the encrusted substrate in several layers.

This order predominantly inhabits shallow seas, although there are a few fresh-water genera. It has existed since the Cambrian, possibly even since the Upper Precambrian.

The genus *Hydractinia* (Eocene-Recent) has a well developed coenosarc (soft body between the individual polyps) and often lives commensally, encrusted on gastropod shells which have been taken over by hermit crabs.

2.   Order: Milleporida

The Milleporida are some of the most important modern reef-builders beside the corals. In fact they used to be classified as corals. They secrete a massive or encrusting calcareous skeleton which contains tubes of varying width, running vertically from the surface and divided by horizontal partitions. The polyps occupy the open part of the tubes. There is a pronounced polymorphism among the polyps: there are gastrozooids (feeding polyps) and dactylozooids (protective polyps). The gastrozooids are short and cylindrical; each individual has four tentacles and occupies a tube with a large lumen, the so-called gastropore. The dactylozooids are long slender mouthless forms with five to seven tentacles armed with nematocysts (Figs. 39 and 40). The individual

Fig. 39. Diagrammatic representation and section of a Recent *Millepora* (×20).

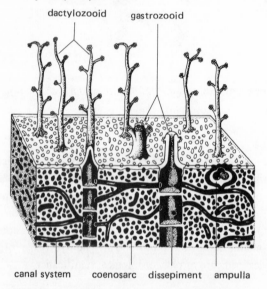

Fig. 40. *Millepora*: (a) shape (×0.5); (b) transverse section (×10); (c) vertical section (×10).

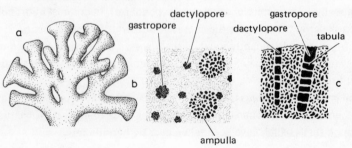

dactylozooids occupy the small narrow, sometimes trumpet-shaped dactylopores which are arranged around the gastropores in a more or less circular fashion and in variable numbers. This arrangement is known as a cyclosystem. The polyps are all connected by a basal canal system in the coenosarc. The small free-swimming medusae are formed in special pores at the surface, the more or less spherical ampullae.

Present-day milleporid species acting as reef-builders do not occur below 50 m, because like the reef corals they live in symbiosis with unicellular green algae. They are rare as fossils, but are known to have existed since the Upper Cretaceous. *Millepora* is a well known Recent and fossil genus with very variable colonial forms depending on its position in its reef habitat.

3.   *Order: Stylasterida*
        The Stylasterida are colonial hydrozoans with a branched or tree-like shape and a hard calcareous skeleton. The coenosarc is further below the surface than in the Milleporida. It contains a network of canals connecting the gastro-zooids, the dactylozooids and the gonophores. The gonophores represent the sexual generation which remains attached to the colony.

Nowadays the Stylasterida are distributed worldwide in shallow waters and down to depths of *c.* 3000 m. The genus *Stylaster* has existed since the Miocene. It is now abundant in warm seas. Its name is derived from the buttresses with radial ribs inside its cyclosystem.

Incertae sedis: Stromatoporida

The Stromatoporida are an extinct group which are usually classified as hydrozoans. Some authors will not agree with this classification, arguing that there is a certain similarity between the stromatopores and Recent sponges of the Class Sclerospongea or calcareous algae, i.e. stromatolites.

In the Silurian and Devonian in particular, the Stromatoporida formed dense encrusting, often reef-forming calcareous masses. The diameter of the colonies varied between less than 1 cm and 1-2 m. The calcareous mass could attain a thickness of 1 m. The calcareous skeleton is divided by horizontal and vertical elements, so that in cross-section it has a multilayered appearance. Vertical pillars, the so-called pilae (singular: pila), and vertical walls of varying length are arranged at right angles to the horizontal, more or less waved calcareous layers, the so-called laminae (singular: lamina) and latilaminae (= a unit consisting of several laminae), which are laid down in parallel with the surface. The laminae and latilaminae can be regarded as evidence for growth periods (Fig. 41).

Fig. 41. *Stromatopora*: (a) full view (x 1); (b) astrorhizae (x 0.8).

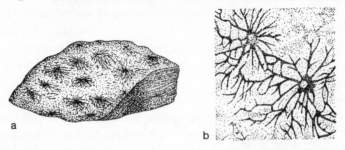

In some forms there are astrorhizae, small branched stellate canals at or just below the surface, which are thought to be zooid stolons, sponges, or the remains of organisms which were (commensally?) associated with the Stromatoporida. Astrorhizae have also been interpreted as the tracks made by boring organisms.

Relatively large round tubes which are divided by horizontal partitions (tabulae) and run vertically from the surface are known as caunopores. When the caunopores have their own walls, they are thought to belong to the tabulate coral *Syringopora* which possibly lived in symbiosis with the Stromatoporida. Small wart-like structures on the surface of the stromatopores are known as mamelons or monticuli. They are often linked with the astrorhizae (Fig. 42).

Fig. 42. *Actinostroma*: (a) full view (x 1); (b) vertical section showing laminae and pillars (x 15).

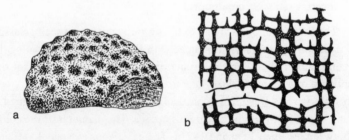

The genera and species of the Stromatoporida can only be identified by the analysis of thin sections. The genera *Actinostroma* (? Cambrian–Lower Carboniferous) and *Stromatopora* (Silurian–Lower Carboniferous; ?Jurassic–?Cretaceous) were distributed worldwide.

3.　　　Class: Anthozoa (corals in the wider sense, sea anemones)

The anthozoans are sessile and exclusively marine Cnidaria that include the corals and sea anemones. They exist only as polyps; there are no medusae.

The polyps occur as solitary forms or in colonies. A tube (stomodaeum) leading from the mouth to the gastral cavity is a characteristic feature. The tentacles are generally arranged round the periphery of the mouth disc, rarely round the mouth opening itself. The polyp wall consists of an external layer (ectoderm), an intermediate layer (mesogloea) and an internal layer (endoderm). The gastro-vascular cavity is divided into radial chambers by four, six, eight or more radial and vertical mesenteries (sarcosepta). Each radial chamber is in contact with one of the hollow retractile tentacles. The mesenteries are usually attached to the stomodaeum; only the incomplete mesenteries project freely into the lumen (Fig. 43).

Fig. 43. Body plan of an hexacoral polyp with an ectoskeleton.

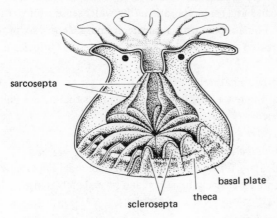

In addition to the radial symmetry imposed by the septa, the anthozoans have bilateral symmetry, which is of importance for classification. Along the mesenteries there are bands of muscle which are bilaterally symmetrical in all anthozoans. In the Hexacorallia they are grouped in pairs, with the exception of the directive septa, whereas in the Octocorallia they are uniformly oriented in one direction on each side. In addition to the muscles, the mesenteries carry other important organs such as the tuberous mesenteric filaments on the inside of the mesenteries which contain the fermentation and resorption cells designed to take up food, and the reproductive cells which are situated in the endoderm. The nervous system forms a network. The anthozoans are dioecious or herm-aphrodite. The zygotes give rise to free-swimming planula larvae which later become sedentary, although the formation of a colony by budding is more common.

Some polyps secrete no hard parts at all, while others produce a calcareous, horny or chitinous skeleton (= corallum) which is either an internal skeleton (usually an axial skeleton) or a compact exoskeleton.

Over 6000 species are known to exist worldwide in the seas today, most of which occur in the Indo-Pacific coral reefs. Fossil anthozoans are found from the Ordovician onwards (Fig. 51).

Taxonomy (overview of the subclasses)

1. Subclass: Cerianthipatharia
   With or without a horny skeleton; simple tentacles; hardly known as fossils
2. Subclass: Octocorallia (= Alcyonaria)
   Polyps with eight mesenteries and eight feathery tentacles; colonial; no great significance as fossils; in existence since the Permian
3. Subclass: Zoantharia (Madreporaria in the wider sense, stony corals)
   Solitary or colonial; some with a calcareous exoskeleton; six protosepta and always paired mesenteries; in existence since the Ordovician

1.   Subclass: Cerianthipatharia
      According to the few fossil finds, the Cerianthipatharia have only existed since the Miocene.

1.   Order: Anthipatharia
      Colonial sedentary forms with a black chitinous axial skeleton with fine thorns. The colonies are extensively branched and several metres high.

2.   Order: Ceriantharia
      Solitary corals up to 50 cm tall, which live in tubes buried in the sediment. Fossilised tubes of the same shape are known as trace fossils; otherwise only Recent forms are known.

2.   Subclass: Octocorallia (Alcyonaria)
      The Octocorallia are exclusively colonial anthozoans with eight feathery tentacles and eight complete mesenteries. The polyps are connected by a system of tubes. They have no calcareous septa. The skeleton is secreted by scleroblasts in the mesogloea which produce a calcareous material or a horny material called gorgonin, thus giving rise to small 0.01–10 mm long spicules or sclerites which may fuse or remain free in the mesogloea. In some forms one axis consists of more or less calcified collagen or alternating calcified internodes and horny nodes (Fig. 44). Six Recent orders are known with about 2500 species, of which some with solid skeletons are important reef-formers. Although fossil representatives from the Permian, Cretaceous and Tertiary are known for four of the orders,

Fig. 44. *Corallium rubrum* with an endoskeleton (*E*). Left, a retracted polyp; on the right, an expanded polyp (× 0.4).

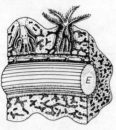

their overall palaeontological importance is minor. The Order Helioporida (blue corals) was originally only known to be fossil. However, the Recent species *Heliopora coerulea* with its solid calcareous skeleton has since been discovered in the Indo-Pacific reef areas.

The following are Recent genera or species of various Octocorallia orders: *Tubipora* (organ-pipe coral), *Alcyonium digitatum* (dead men's fingers), *Gorgonia flabellum* (syn. *Rhipidogorgia flabellum*, sea fan), *Pennatula* (sea pen) and *Corallium rubrum* (precious coral) (Figs. 44 and 45).

Isis and *Moltkia* of the Upper Cretaceous are fossil genera of the Octocorallia (Fig. 45).

Fig. 45. Recent and fossil octocorals. *Isis* and *Moltkia* (× 1); *Pennatula* (× 0.3); *Heliopora* (× 3); *Tubipora* (× 0.5).

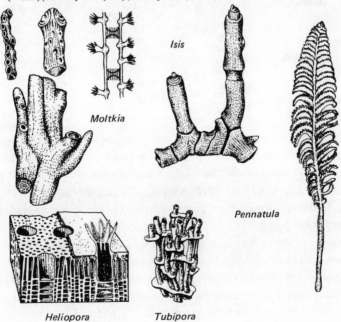

Moltkia

Isis

Pennatula

Heliopora   Tubipora

3.   Subclass: Zoantharia (Madreporaria, stony corals)

Zoantharia are solitary or colonial Cnidaria. The individuals (polyps) consist of the pedal disc, which can secrete a calcareous basal plate, the body and the oral disc. The lateral body walls of a polyp are known as the pallium.

In many zoantharians the exoskeleton of calcium carbonate is secreted by the ectoderm. The skeleton of a single polyp is called a corallite (cup), while the skeleton of an entire colony is a corallum.

The skeleton always has a fibrous prismatic structure. The fibres are arranged in parallel with each other (usually oblique to the longitudinal axis of the skeletal element in question) or as trabeculae (radially around linear centres). The peripheral border of the coral polyp, the polyp wall, may have one of three different basic shapes: the eutheca, epitheca (Fig. 46) or pseudotheca:

1. The eutheca is a true theca which is secreted in an annular fold of the pedal disc. It has a dark midline from which the fibres point obliquely upwards on both sides. Euthecae occur predominantly in the Cyclocorallia.

Fig. 46. Schematic diagram of the skeletal elements of a polyp.

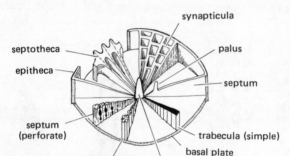

2. The epitheca is a calcareous tube with horizontal wrinkles, which is secreted by the external surface of the pallium (= polyp mantle), i.e. it is secreted by only one surface. The epitheca is thus a direct continuation of the basal plate, which in the Cyclocorallia is present externally in addition to the eutheca. In the Pterocorallia, on the other hand, the epitheca represents the only envelope for the polyp.

3. The pseudotheca may arise from quite different structures, e.g. from thickened septal bases touching each other (septotheca, septothecal wall), from loosely connected (dissepimental wall) or directly adjacent dissepiments (see below; parathecal wall) consisting of one or more rows of synapticulae (synapticulotheca). Synapticulae, which are particularly pronounced in the Cyclo-

corallia, are rods emanating from neighbouring septa and growing towards each other. On penetrating the mesenteries by the formation of a connecting piece which has its own centre of crystallisation, the synapticulae unite in the interseptal spaces.

The septa inside the polyp are also formed in the basal part of the polyp, namely in the radial folds of the pedal disc. While the polyp grows, sclerodermites are formed. These are calcareous fibre bundles which emanate from a centre of calcification. In the septa the sclerodermites are arranged in alternating rows; they also form the trabeculae (Fig. 46).

The most common arrangement of the trabeculae in the septa is parallel and oblique from the bottom outside to the top inside or fanning out towards the top. The septal surfaces are usually granulated, occasionally smooth; the septal edges are waved, serrated, rippled or smooth. On the external wall of the corallite continuations of the septa are often visible as costae (ribs).

The ontogeny of a polyp can be followed through a series of transverse sections. The sequence of the septal layout is of taxonomic importance.

Inside the cup, those septal ends which are bowed sideways may fuse and create an internal tube, the aulos, which separates the inner and outer parts of the tabulae (see below) in certain pterocorals, for example (Fig. 52*B*, 7).

In many corals there is an axial structure, the columella, which is derived from the fusion of separated axial septal ends with an independent nucleus growing upwards (Fig. 52*A*, 3). If the latter is absent, the structure is known as a pseudocolumella.

The pali (singular: palus) in the Cyclocorallia are more or less rod-shaped vertical elements between the columella and the septa. They represent the separated innermost parts of septa of certain intercalatory cycles, i.e. they are secondary septal structures (Fig. 46).

During its development the polyp is capable of secreting successive skeletal elements in its basal portion. The following elements are distinguished (Fig. 47):

Fig. 47. Schematic diagram of the development of dissepiments and tabulae in pterocorals.

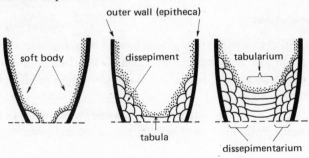

1. The tabula (table), a more or less horizontal thin plate which passes through the whole interior of a polyp or through its central part. In the latter case the part of the polyp occupied by the tabulae is known as the tabularium. Tabulae may be complete, in which case they extend from wall to wall or across the whole tabularium, or incomplete, in which case they rest on different bases and only touch the wall or dissepimentarium (see below) on one side. Incomplete tabulae are also known as tabellae (singular: tabella).
2. Dissepiments, plate-like curved skeletal elements which form the boundaries to bladder-shaped cavities. They are characteristic in certain Pterocorallia genera and are usually restricted to the peripheral portion of the interseptal areas. The area occupied by the dissepiments is known as the dissepimentarium and is in contact with the tabularium along its periphery. The dissepiments originate either from the pedal disc itself or from parts of the pedal disc and the pallium. They originate post-septally (Fig. 47).

The skeletal elements can thus be classified in the following manner:

1. radial skeletal elements: septa, costae and pali;
2. tangential skeletal elements: eutheca, pseudotheca and epitheca;
3. horizontal skeletal elements: tabulae, dissepiments and synapticulae;
4. central skeletal elements: columella and aulos.

The skeletal material connecting the individual corallites within a colony, which is secreted by the coenosarc (soft body between the individual polyps), is known as coenenchyme.

For the classification of corals the shape of the corallum and the polyps is of little importance since in these sessile animals this is very dependent on the ecological conditions prevailing in their habitat. Several descriptive morphological terms have been coined for the various coral shapes. Convergent and parallel evolution are of importance. The evolution of the septa is more important for classification than that of the cup.

Reproduction

In addition to sexual reproduction, asexual reproduction by budding plays an important role. In asexual reproduction the new individuals invariably remain in contact with each other and with the parent polyps and thus form coral reefs or colonies.

If budding occurs within the cup or calice, it is known as internal or intra-tentacular budding. With this type of budding the mouth and tentacles of the

new polyp are formed within the oral disc of the parent polyp. At the same
time individual septa enlarge and link up within the calice until a new polyp is
enclosed. In addition to septal budding there is also tabular budding in which the
tabulae inside the calice arch upwards to the extent that they form the external
walls of new corallites. Rejuvenescence is a special type of internal budding in
which the parent calice gives rise to only one bud. This enlarges until it fills out
the parent calice. This type of budding is repeated many times until eventually
a column is formed which consists of a series of cups, one on top of or inside the
other.

In external or extratentacular budding the polyps are usually formed from
the side of the parent polyp. In addition to the mechanism of lateral budding,
new polyps can also develop from the coenosarc (coenosarc budding) or, less
often, from stolons produced by the parent polyp (stolonial budding).

Ecology of the corals

Corals are exclusively marine organisms which predominantly inhabit
shallow sea areas and which are occasionally reported down to depths of 6000 m.
Most corals are reef-formers (hermatypic) and prefer warm shallow sea areas with
clear water and depths not exceeding 80-90 m. For reef corals the optimum living
conditions are provided in the well lit (photic) region above 35 m at a tropical
water temperature of 25-28 °C and a salt content of $35-40^0/_{00}$ (Fig. 48). Their
occurrence at this depth is linked with their symbiotic relations with unicellular
green algae (Zooxanthellae: *Gymnodinium microadriaticum*) which require
sunlight for photosynthesis. The corals use the oxygen which is liberated in the
photosynthetic process. In return, the Zooxanthellae are protected by the coral
tissues and have access to the carbon dioxide which is vital to them and which the
polyps liberate during respiration. By assimilating the carbon dioxide they also
facilitate the secretion of calcium carbonate. The symbiotic Zooxanthellae are
thus important in oxygen and calcium exchange (Fig. 49). Colonies of some
genera, e.g. *Lophohelia* and *Dendrophyllia*, are also found in colder waters and
at greater depths (Atlantic coasts of France, Norway and Greenland). They are
ahermatypic and do not act as hosts to Zooxanthellae.

Most of the present-day reef corals live between the latitudes 28° North and
28° South. The largest Recent reef in this equatorial reef belt is the Great
Barrier Reef off the east coast of Australia. It is about 3000 km long and in some
places 100-300 km wide. Many observations indicate that the fossil coral reefs
were also created in relatively warm water and that they also formed a worldwide,
more or less equatorial, reef belt.

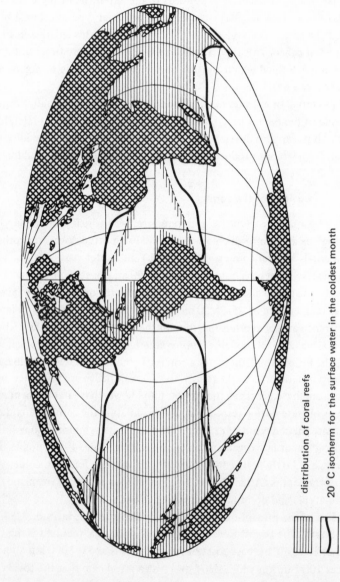

Fig. 48. Distribution of Recent coral reefs.

distribution of coral reefs

20°C isotherm for the surface water in the coldest month

Fig. 49. Model of calcium carbonate secretion in stony corals.

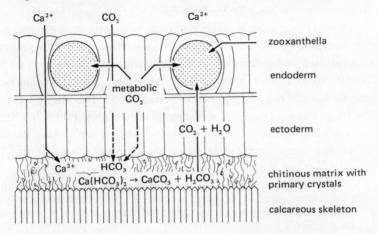

Summary of the palaeontologically significant orders of the Subclass Zoantharia:

1. Order: Pterocorallia (Rugosa, Tetracorallia). Occurrence: Ordovician–Permian
2. Order: Cyclocorallia (Scleractinia, Hexacorallia). Occurrence: Triassic–Recent
3. Order: Heterocorallia. A small group restricted to the Lower Carboniferous
4. Order: Tabulata. Occurrence: Ordovician–Permian

1.   *Order: Pterocorallia* (syn. Rugosa, Tetracorallia)

This order (Fig. 50*A*) was named after the feather-like arrangement of the septa (Pterocorallia), the horizontal wrinkles (rugae) on the external wall

Fig. 50*A*. Morphology of the pterocorallian corals: (a) longitudinal section; (b) transverse section.

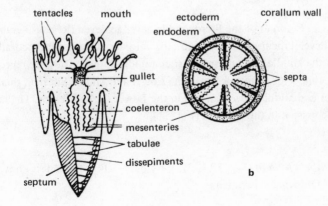

(Rugosa), or the arrangement of the septa into four quadrants (Tetracorallia).

The forms in this group are predominantly solitary, occasionally colonial.

The layout of the septa is always bilaterally symmetrical. The Pterocorallia are characterised by a special type of septal intercalation. After the formation of the six protosepta (first-formed septa), subsequent septa (metasepta) are formed only in the four quadrants between cardinal and lateral (alar) septa and between lateral and counter-lateral septa (hence 'Tetracorallia'). This insertion of septa in a pinnate arrangement proceeds according to 'Kunth's law' (1869/70). Short minor septa may be inserted at a later stage in the gaps between the counter-septa and the counter-lateral septa, which remain free of metasepta (Fig. 50*B*).

Fig. 50*B*. Insertion of septa in the Pterocorallia. C, cardinal septum; K, counter-septum; L, lateral septum; KL, counter-lateral septum; I, II, III, metasepta in the sequence of their insertion; M, minor septum.

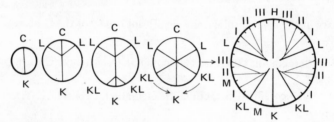

The Pterocorallia are also characterised by the abundance of tabulae. The tabulae either traverse the entire corallite or, if a marginarium has been developed from dissepiments (dissepimentarium), they are restricted to the tabularium. The development of an epitheca is also characteristic of the Pterocorallia. A depression in the tabulae, usually in the region of the cardinal septum, is known as the fossula.

As the development of the skeletal elements changes considerably during the course of ontogeny, numerous modifications are visible in horizontal and longitudinal sections.

Occurrence: Colonies of the Pterocorallia never achieved the considerable spread and development of the Cyclocorallia colonies. The first Pterocorallia appeared in the Middle Ordovician and then rapidly spread. They are represented for the first time with many species in the Lower and Middle Silurian, then again in the Lower and Middle Devonian and in the Lower Carboniferous. The last Pterocorallia existed in the Permian (Fig. 51).

Examples:

Lambeophyllum (Fig. 52*A*, 2): Very deep calice; the oldest known genus. Occurrence: Ordovician–Permian.

Fig. 51. Chronological distribution of the Pterocorallia and Cyclocorallia.

Quaternary	*Acropora* *Flabellum*
Tertiary	*Porites*
Cretaceous	*Cyclolites* *Thecosmilia*
Jurassic	*Parasmilia*
	Caryophyllum Cyclocorallia
Triassic	(Scleractinia)
Permian	
	Hexagonaria
Carboniferous	*Clisiophyllum, Lonsdaleia*
	Zaphrentis, Calceola
Devonian	*Streptelasma*
Silurian	*Goniophyllum*
	Lambeophyllum
Ordovician	Pterocorallia
	(Rugosa)
Cambrian	

Fig. 52*A*. Pterocorallia: (1) *Zaphrentis* (×1); (2) *Lambeophyllum* (×1); (3) *Cyathaxonia* (×1); (4) *Streptelasma* (×1); (5) *Cystiphyllum* (×0.5); (6) *Acervularia* (×1).

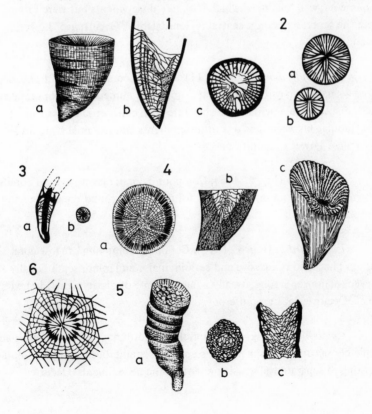

Cyathaxonia (Fig. 52*A*, 3): Small, with columella independent of the septa, long minor septa, complete tabulae. Occurrence: Lower Carboniferous.

Streptelasma (Fig. 52*A*, 4): Short minor septa, marginarium as a septal stereozone (= thickened skeletal zone), no dissepiments; narrow loose axial structure consisting of the axial lobes of the cardinal septa. Occurrence: Middle Ordovician–Middle Silurian.

Zaphrentis (Fig. 52*A*, 1): Solitary or colonial; narrow dissepimentarium; the cardinal septa meet axially; central swelling consisting of domed tabulae. Occurrence: Devonian.

Clisiophyllum (Fig. 52*B*, 8): Usually solitary; with extensive dissepimental zone; incomplete conical tabulae; axial structure; small open fossula; short thickened median plate (= aulos). Occurrence: Lower Carboniferous.

Lonsdaleia (Fig. 52*B*, 7): Thick columella; central tabulae and marginal dissepiments; wall 'bladders' which look like dissepiments but were formed before the septa; septa only central (= lonsdaleoid). Occurrence: Lower Carboniferous.

Dibunophyllum (Fig. 52*B*, 11): Solitary; corallite cylindrical except at the base; small open cardinal fossula; major septa numerous, minor septa irregular, closely spaced dissepiments; axial structure with concentric and radially arranged plates; tabulae between the zone of dissepiments and the axial structure. Occurrence: Lower Carboniferous.

Acervularia (Fig. 52*A*, 6): Compound coral; massive cerioid corallum; inner ends of the septa thickened and forming a wall round the axial region. Occurrence: Silurian.

Lithostrotion (Fig. 52*A*, 12): Common, compound, rugose coral; corallum phaceloid or massive and cerioid; major and minor septa radially arranged, major septa long; usually a narrow zone of dissepiments; arched tabulae. Occurrence: Carboniferous.

Cystiphyllum (Fig. 52*A*, 5): Solitary horn-shaped coral; interior almost completely occupied by spherical dissepiments; numerous incomplete septa consisting of separate trabeculae present on the dissepiments. Occurrence: Silurian.

Fig. 52B. Pterocorallia: (7) *Lonsdaleia* (×1): (8) *Clisiophyllum* (×0.5);
(9) *Goniophyllum* (×0.5); (10) *Calceola* (×0.5); (11) *Dibunophyllum*
(×1); (12) *Lithostrotion* (a: ×3; b, c: ×1).

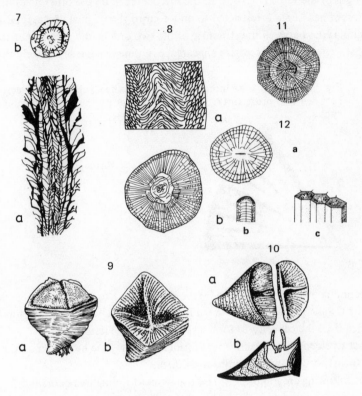

Calceola (Fig. 52B, 10): Solitary slipper-shaped coral with triangular
base; deep calice; one-part hemispherical lid; short septa and tabulae; dissepiments
more or less bladder-shaped. Occurrence: Lower to Middle Devonian. *Calceola
sandalina* is an index fossil for the Middle Devonian.

Goniophyllum (Fig. 52B, 9): Solitary pyramidal coral with a more or
less curved point which may have root-like outgrowths; deep calice with a quadri-
partite lid; thick leaf-shaped septa; vesicular tissue; numerous more or less
thickened tabulae. Occurrence: Silurian.

2.   *Order: Cyclocorallia*
        This order is characterised by the cyclic insertion of septa. Between the
six protosepta, which arise in the same way as those of the Pterocorallia, six
further, second-order septa are formed. Between these and the protosepta,

12 third-order septa are formed, then 24 fourth-order septa, etc. The basic unit of six is thus maintained, while the cyclical arrangement is usually expressed in the length of the septa (Fig. 53). In the Pterocorallia, on the other hand, the addition of new septa is restricted to only four of the original protoseptal spaces. The similarities between the structure of the two orders, the presence of transitional stages and their successive appearance in time make it seem likely that the

Fig. 53. Model of the insertion of septa in the Cyclocorallia. Sector between two protosepta. The numbers indicate the number of septal cycles.

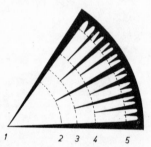

Cyclocorallia originated from the Pterocorallia. It is also possible, though unlikely, that the Cyclocorallia already existed in the Palaeozoic without a skeleton and only acquired the capacity to form a skeleton in the Triassic.

Most Cyclocorallia are active reef-formers, i.e. they are hermatypic. Ahermatypic forms occur down to depths of 6000 m.

The following are important present-day reef-forming Cyclocorallia:

Acropora (Fig. 54*A*, 3): Pacific Ocean; important reef-former; 200 Recent species (40% of the living Cyclocorallia); skeleton very light and porous, branched; calices usually large at the end of branches and smaller laterally; no columella. Occurrence: Eocene–Recent.

Porites (Fig. 54*A*, 2): Colonial; massive and sometimes furnished with blunt lateral branches or encrusting; intratentacular budding. The calices (up to 2 mm in diameter) are usually close together. The septa are regularly perforate and consist of three to four more or less vertical trabeculae. Trabecular columella with a ring of pali; distributed worldwide; important reef-formers with about 100 Recent species. Occurrence: Jurassic–Recent.

Cyclolites (Fig. 54*A*, 1): Solitary, disc-shaped, unattached; corallum domed at the top and flat at the bottom with wrinkled epitheca (concentric

Fig. 54*A*. Cyclocorallia: (1) *Cyclolites* (× 0.33); (2) *Porites* (a: × 0.2; b: × 6); (3) *Acropora* (a: × 0.125; b: × 0.5; c: × 3); (4) *Fungia* (× 0.33).

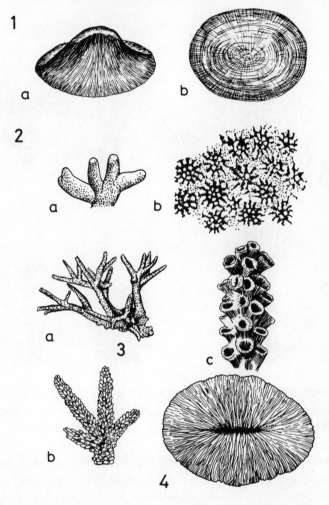

rings); septa thin and perforate. The structure of the septal apparatus is homeomorphic with that of *Fungia*. Occurrence: Cretaceous–Eocene.

Thecosmilia (Fig. 54*B*, 5): Colonial; calices touching each other, distal end free; numerous granulated or ribbed septa; no columella; long-lived with many species; included is the genus *Montlivaltia*; distributed worldwide. Occurrence: Triassic–Eocene.

Fig. 54B. Cyclocorallia: (5) *Thecosmilia* (× 0.33); (6) *Madrepora* (× 0.5); (7) *Caryophyllia* (× 1); (8) *Flabellum* (× 0.5); (9) *Parasmilia* (× 1).

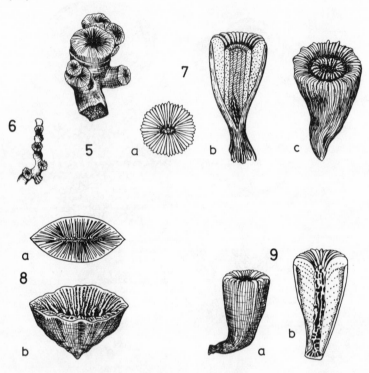

Flabellum (Fig. 54B, 8): Solitary, usually a flattened horn shape; septa smooth or granulated consisting of fan-shaped system of simple trabeculae; no dissepiments or pali; distributed worldwide. Occurrence: Eocene–Recent.

Caryophyllia (Fig. 54B, 7): Usually solitary and horn-shaped; extra-tentacular budding; numerous septa, either smooth or finely granulated; columella with more or less lobe-shaped pali; worldwide distribution. Occurrence: Malm–Recent.

Parasmilia (Fig. 54B, 9): Solitary, flat cone-shaped coral grown up from a very small base; dissepiments deep inside the calice; spongy columella; world-wide distribution. Occurrence: Cretaceous–Recent.

Fungia (Fig. 54A, 4): Large mushroom-shaped solitary coral with numerous septa connected by synapticulae; theca absent; tropical Indo-Pacific with 46 species. Occurrence: Miocene–Recent.

3.   *Order: Heterocorallia*

The zoantharian order of the Heterocorallia is restricted to the Lower Carboniferous (Viséan) with only two genera from Germany, Scotland and Japan.

The four protosepta form a simple axial cross; the other septa are formed by peripheral division of the protosepta (Fig. 55).

Fig. 55. Model for the multiplication of septa in the Heterocorallia (*Heterophyllia*). The circles indicate successive growth stages.

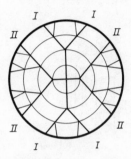

Hexaphyllia (Fig. 56*A*, 1) and *Heterophyllia*: Narrow and cylindrical with a diameter of 5–15 mm and a length of up to 50 cm; arrangement of septa initially like that of the Pterocorallia, then bifurcation, giving rise to four fossulae in the areas between the primary septa.

The change from the Pterocorallia occurred by proterogenesis, i.e. by a mutation acting early in ontogeny which became integrated in the normal course of development and considerably changed ontogeny.

4.   *Order: Tabulata*

All members of this order are exclusively colonial corals with numerous tabulae and underdeveloped septa. The septa are usually only present as rows of spines. The corallites are usually slender (0.2–5 mm in diameter) and tube-shaped. The walls are perforated by numerous rows of pores. The shape of the corallum is again very diverse; there are massive tuberous shapes as well as branched forms. Bulbous forms can reach a diameter of 1–50 cm.

The origin of the tabulate corals is still uncertain. There is no doubt that they existed from the Middle Ordovician to the Permian, which means that they are the same age as forms of the Order Pterocorallia. The taxonomic status of the Tabulata is also disputed; they probably represent a heterogeneous order.

Examples:

Favosites (Fig. 56*A*, 2): Mainly massive colonies with more or less pronounced polygonal corallites. Walls with short septa which are present either

Fig. 56*A*. Heterocorallia: (1) *Hexaphyllia* (×6). Tabulate corals:
(2) *Favosites* (a: ×0.5; b: ×3); (3) *Pleurodictyum* with *Hicetes* (×1);
(4) *Syringopora* (a: ×0.5; b: ×2).

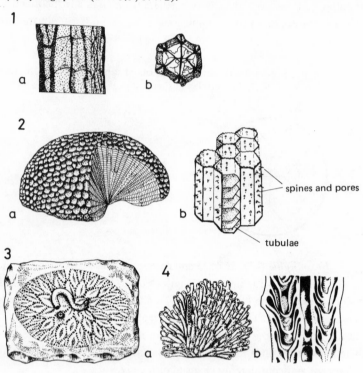

as ridges or thorns. Walls always have pores and the tabulae are always close
together. Distributed worldwide from the Upper Ordovician to Middle Devonian.
Indications that the favositids might belong to the Sclerospongea are being
discussed at the moment.

Pleurodictyum (Fig. 56*A*, 3): Disc-shaped to hemispherical, more or
less circular colonies which often enclose a U-shaped worm (*Hicetes*) (synoecism
or parasitism?). The colonies are often preserved as a *Steinkern* (mould). The
walls are thick and perforated by large pores; the septa are reduced to rows of
thorns; there are no tabulae. Distributed worldwide in the Lower Devonian.

Pleurodictyum problematicum is characteristic of the sandy facies of the
Lower Devonian of Central Europe.

Halysites (Fig. 56*B*, 6): This genus has long cylindrical tubes which
are fused along their narrow edge. The corallites have tabulae. This so-called

Fig. 56*B*. Tabulate corals: (5) *Aulopora* (a: × 0.5; b: × 1.5); (6) *Halysites* (a: × 0.7; b, c: × 3); (7) *Heliolites* (a: × 0.5; b: × 10; c: × 4). Tab., tabulae; Cor., corallite; M.Cor., microcorallite.

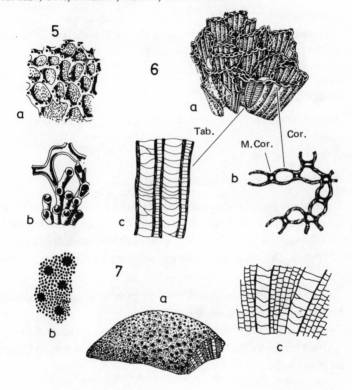

chain coral is distributed worldwide from the Ordovician to the Silurian; it is absent only in South America.

Aulopora (Fig. 56*B*, 5): Colonies branched in a net-like fashion, encrusting the substrate with their underside. Homeomorphic with the bryozoan genus *Stromatopora*. Worldwide distribution in the Devonian.

Syringopora (Fig. 56*A*, 4): Colonies consist of thin cylindrical, slightly curved corallites which are linked by stolons. Tabulae close together. Worldwide distribution, with the exception of South America, from the Silurian to the Upper Carboniferous. *Syringopora* also occurs endobiotically (as '*Caunopora*') in the stromatopores.

The two genera *Chaetetes* and *Heliolites* are often classified as Tabulata, but it seems very unlikely that they are. *Chaetetes* has recently been assigned to the

Porifera (Class Sclerospongea), and *Heliolites* is listed as a genus *incertae sedis* here.

Heliolites (Fig. 56*B*, 7): Massive hemispherical colonies. The corallites have 12 septa which are sometimes present as thorns. Distributed worldwide from the Upper Silurian to the Middle Devonian.

2.   Subgroup: Protostomia

Animals in which the larval mouth is often retained and becomes the actual mouth.

9.   *Phylum: Mollusca* (soft-bodied animals)

Molluscs are protostomes with unsegmented solid bodies; only the most primitive ones are segmented. Their phylogenetic derivation is not known for certain; there is evidence of a relationship with both the annelids and the turbellarians (flatworms). The molluscan body can generally be divided into four parts (Fig. 57):

1. The more or less distinct head with the mouth, tentacles and eyes (the head is absent in the bivalves).
2. The muscular ventral foot, a local thickening of the muscular envelope; it serves primarily for locomotion, but may be differentiated in many different ways.

Fig. 57. Fundamental patterns of molluscan organisation: (a) cephalopod, (b) gastropod, (c) scaphopod, (d) lamellibranch.

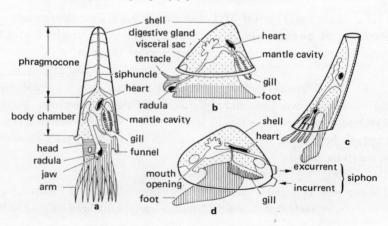

3. The dorsal visceral mass containing the intestines, kidneys and reproductive system.
4. The mantle (pallium), a dorsal fold of skin. The mantle secretes the shell and forms the boundary of the mantle cavity (= respiratory cavity) which contains the gills ('ctenidia', comb-shaped gills).

The head, foot or mantle may be reduced to the point where they are virtually absent. The nervous system with its ganglia, on the other hand, is always present and may achieve an extraordinarily high level of differentiation (in some cephalopods the brain is encased in cartilage).

Haemocyanin serves as the blood pigment in most cases (oxygen-carrying capacity in *Loligo* is 4.2%, in *Helix* 2%; human haemoglobin, on the other hand, has a capacity of 20%).

Two structures form the foundation of mollusc organisation:

1. The radula: a characteristic molluscan organ which has the function of breaking up or swallowing food. It forms the basis for molluscan versatility, but is absent in the bivalves.
2. The shell: the presence of an inflexible shell necessitates special organs of locomotion such as the foot or the funnel. The shell is single in scaphopods, gastropods and cephalopods, but consists of two parts in the lamellibranchs and of eight parts in the amphineurans.

With approximately 128 000 Recent species, the molluscs are the most diverse phylum after the arthropods. They inhabit oceans, fresh waters and terrestrial habitats up to great heights.

Representatives of the cephalopod genus *Architeuthis* grow to the largest size (with arms exceeding 20 m they are the largest invertebrate animals). The smallest molluscs are only a few millimetres long.

Structure of the shell

The molluscan shell material is usually made up of two components, the proportions of which can vary:

1. The periostracum, which consists of dense organic material (conchiolin, a scleroprotein composed of a number of amino acids) and is formed by the part of the mantle projecting beyond the shell.
2. Mineral shell material, which consists either of calcite or aragonite or both in varying proportions. The mineral substance contains varying proportions of organic substance arranged in networks or frameworks or finely spread out in the shell material ('amorphous').

The first shell chamber is known as the protoconch.
The shell forms the basis for classification, and accordingly two subphyla are

distinguished (Fig. 58):

    1. Subphylum: Amphineura (Aculifera, chitons)
        1. Class: Polyplacophora (chitons in the narrower sense)
        2. Class: Caudofauveata
        3. Class: Solenogastres (Aplacophora)
    2. Subphylum: Conchifera (molluscs with shells consisting of one or two parts; Fig. 58)

Fig. 58. The Recent classes of the Mollusca.

	Class	Diagrammatic representation	Number of species	
			Recent	Fossil
Aculifera	Polyplacophora		1000	100
	Caudofoveata	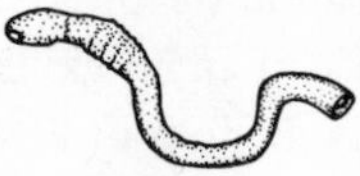	55	–
	Solenogastres		110	–
Conchifera	Monoplacophora	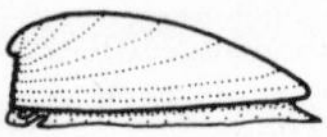	4	50
	Gastropoda		105 000	15 000
	Scaphopoda	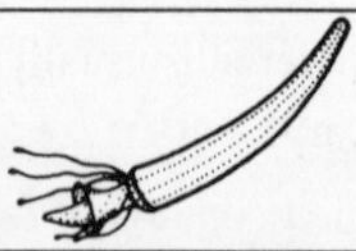	350	200
	Lamellibranchia	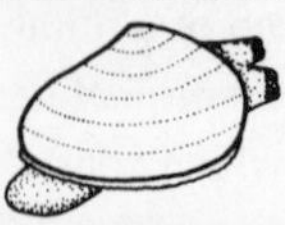	20 000	15 000
	Cephalopoda		730	10 500

1. Class: Monoplacophora
2. Class: Gastropoda (snails)
3. Class: Scaphopoda (tusk-shells)
4. Class: Rostroconchia (extinct)
5. Class: Lamellibranchia (bivalves)
6. Class: Cephalopoda

There are also two extinct classes of very small shelled animals which are thought to have been molluscs:

7. Class: Cricoconarida (tentaculitids among others)
8. Class: Calyptoptomatida (hyolithids among others)

1.   ***Subphylum: Amphineura*** (chitons)

Of the three externally very different classes united in this subphylum only the first has fossil representatives.

1.   Class: Polyplacophora (chitons – Placophora, Loricata – in the narrower sense)

The polyplacophoran body has no eyes, tentacles or nerve ganglia, but it has light-sensitive cells (aesthetes) on the shell surface, a subradular organ (chemical sense organ for taste and smell) as well as numerous gills in the lateral mantle grooves. The chitons are usually 2.5–7.5 cm long. Their radula has 17 teeth per transverse row and is very long. The shell consists of a series of eight (very occasionally seven) calcareous plates (Fig. 59).

Fig. 59. Polyplacophoran (chiton): (a) ventral view, (b) dorsal view (×1).

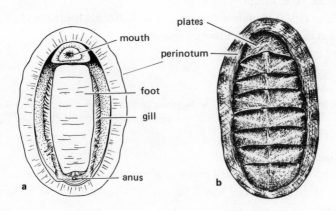

In contrast to the shell material of other molluscs, that of the chitons consists of three layers:

1. The extremely thin periostracum made up of organic substance.
2. Underneath, the tegmentum which is penetrated by pores. It consists largely of organic substance with deposits of aragonite crystallites in the form of spherulites.
3. Underneath the tegmentum lies the aragonitic hypostracum with its cross-lamellar structure. Embedded in the hypostracum is the articulamentum made up of areas of spherulites. The articulamentum was formerly regarded as an independent layer of shell material. It makes up the apophyses in particular. A myostracum made up of large prisms of complicated structure is found at the muscle attachment sites.

The plates are encircled by the girdle (perinotum), which is covered with calcareous spines or scales.

The plates of Palaeozoic forms hardly overlapped and were sometimes thick and solid. 'Modern' forms with apophyses are known since the Upper Palaeozoic. Fossil finds are rare and tend to consist only of isolated plates, so that the amphineurans have no great palaeontological significance.

Amphineurans prefer to live on hard substrates which they adhere to with their broad foot. They occur down to depths of 5000 m. Their diet consists of algae, bryozoans, hydroids, etc., with each species specialising in a certain diet.

1.   Order: Paleoloricata
Primitive forms with thick massive valves, not overlapping; no articulamentum.
Occurrence: Cambrian–Cretaceous.

2.   Order: Neoloricata
Valve structure more differentiated, with articulamentum and insertion plates, except in the most primitive representatives.
Occurrence: Carboniferous–Recent.

2.   Class: Caudofoveata
Worm-like; animals living in burrows in the sea-bottom. No known fossil representatives.

3.   Class: Solenogastres
Solenogastres are a small group of worm-like molluscs without a calcareous shell; instead they have a thick tough cuticle. There are no known fossil representatives.

2.   **Subphylum: Conchifera**

This subphylum includes the great majority of molluscs (over 126 000 Recent species compared with 1150 Recent amphineuran species).

1.   Class: Monoplacophora

When this class was established in 1952, it contained only a small number of Palaeozoic fossils. In the same year, a Recent representative was discovered by the Danish research ship *Galathea* west of Central America at a depth of 3570 m. It was later (1957) called *Neopilina galatheae* by Lemche.

The characteristic features of *Neopilina* (Fig. 60, 3) are: almost bilaterally symmetrical structure with internal metamerism, a large uniform, slightly

Fig. 60. Monoplacophora: (1) *Tryblidium* (× 1); (2) *Scenella* (a: × 2.5; b: × 2); (3) *Neopilina* (a: apical part of larval shell × 2.5; b: dorsal view × 0.8; c: ventral view × 1).

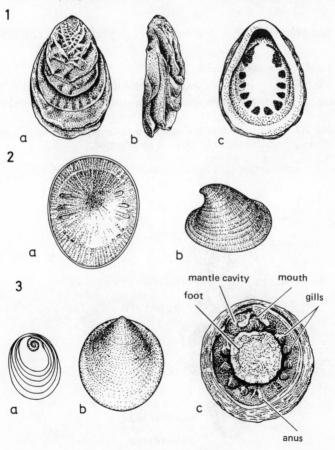

arched thin dorsal shell, an asymmetrical protoconch. Metameric nephridia
(organs of excretion) terminate in the mantle canal which contains five pairs
of feathery gills. Numerous muscles connect the soft body to the shell.

Similar cap-shaped mollusc shells with several pairs of muscle scars found
in the Palaeozoic had previously been assigned to the gastropods. The characteristics of the new Class Monoplacophora are thus: a uniform cap-shaped shell
with several pairs of muscle scars inside the shell.

Examples:

Scenella (Fig. 60. 2): Low hat to cap shape; six or seven pairs of muscle
scars. Occurrence: Cambrian, mainly in Canada.

Tryblidium (Fig. 60. 1): Flat spoon-shaped; anterior muscle scars
stronger; wide lamellae on the outside; very fine perforations in the shell.
Occurrence: Silurian of Gotland.

Neopilina (Fig. 60.3): See description above. Occurrence: Recent.

2.   Class: Gastropoda (snails)
Gastropods are molluscs with a head, foot, visceral sac and mantle.
The mantle can secrete a uniform spiral or bowl-shaped calcareous shell (Fig. 61).

Fig. 61. Descriptive terminology for a gastropod shell, using a genus
of the Neogastropoda (*Latirus*) as an example.

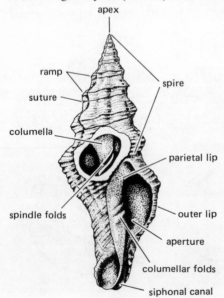

With about 105 000 Recent species the gastropods are the most diverse and most widely distributed class of molluscs (Table 4, p. 42). Originally they were exclusively marine, but at least from the Carboniferous onwards they also occurred in fresh waters and on land. Although they usually inhabit shallower waters, they are also found at depths of up to 5000 m and at heights of 6000 m. They are vegetarian, carnivorous or omnivorous.

The head with the mouth, eyes and one or two pairs of tentacles is quite distinct. Either side of the radula, which is situated on the floor of the mouth, there are sometimes two small horny plates, the so-called jaws. A salivary gland in the gullet secretes a sometimes poisonous digestive juice (Fig. 62).

Fig. 62. Longitudinal section of a gastropod head showing the position of the radula. F, foot; J, jaw; MO, mouth opening; OE, oesophagus; RP, radula pad; RS, radula sac; SG, salivary gland.

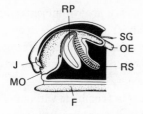

The interaction of longitudinal and transverse muscles in the muscular foot makes it possible for the animal to glide along on uneven surfaces. The foot can be modified into wing-like organs for swimming.

The mantle cavity contains the comb-like gills (ctenidia), or, in the case of pulmonates, the surface of the mantle is differentiated into 'lungs'. The osphradium is a special sensory organ near the gills which can bring about the closure of the mantle cavity (if the water is very dirty, for example). The siphon, a tube-like extension of the mantle, can facilitate respiration on muddy substrates.

Torsion is the most characteristic feature of the gastropods. It involves the rotation of the mantle and shell, including the enclosed soft parts, anticlockwise by 180° with respect to the head and foot. Torsion may occur at an early stage. For the adult animal the consequences are as follows: the anus and mantle cavity with the gills are twisted forwards or sideways (the gills then come to lie in front of the heart, hence the name Prosobranchia); at the same time the nervous system is twisted into the 'streptoneural' position (so-called chiastoneury, as opposed to orthoneury where the nerve cords are not twisted). Products of excretion and reproduction will thus exit above the head, but a median slit along the edge of the shell aperture channels the current dorsally.

Primitive prosobranchs (archaeogastropods) release their eggs and sperm into the water (external fertilisation), so that their development is tied to the sea. All other prosobranchs have internal fertilisation which is possible both on land and in the water. This step made the colonisation of rivers and land possible as well as processes such as ovovivipary (hatching of the larvae from the eggs inside the maternal body and subsequent 'birth').

Hard parts

The shell material consists of the external periostracum layer of conchiolin and the mineralised layer with a greater or lesser amount of organic substance. The mineral component may be calcitic or aragonitic or both.

The periostracum is hidden in those parts of the shell covered by folded-back parts of the mantle. The large proportion of aragonite in the shell material often causes fossilisation only as moulds (*Steinkerne*), even in environments where calcitic shells (e.g. of brachiopods or many bivalves) have been preserved.

The embryonal whorls (= protoconch; later whorls constitute the teleoconch) may be homeostrophic or heterostrophic (if they do not coil in the same axis as the teleoconch).

The aperture (peristome) may be holostomatous (complete without any particular features) or siphonostomatous (with a siphonal canal) (Fig. 61).

The slit-band (selenizone) is a narrow groove on the outside of the whorls of the shells of some gastropods, especially the archaeogastropods, which is constructed in a different way from the rest of the shell. Its crescent-shaped growth lines are called lunules. As the anal canal moves forward, the slit-band progressively closes the slit in the shell which is held open by the former.

Most prosobranchs have a lid (operculum) which closes the aperture. The operculum is secreted by the posterior part of the foot and is usually horny and occasionally calcareous. In dextrally coiled (in the clockwise direction when seen from above: Fig. 63) gastropods the operculum is unrolled anticlockwise,

Fig. 63. Terrestrial snail *Bulimus*: (a) sinistrally coiled, (b) dextrally coiled (×0.5).

a   b

so that it is possible to ascertain the direction of coiling even in extinct forms. In some pulmonates, on the other hand, there is an epiphragm ('winter operculum'), a lid consisting of slime and calcareous material which is used in periods of hibernation (in winter or in droughts) and then discarded again.

Radula

In Recent prosobranchs the radula is used for more detailed classification; it is not known as a fossilised structure. Depending on the number and arrangement of teeth in each of the numerous transverse rows, it is possible to distinguish different types of radula. The rhachidian tooth, usually a specially differentiated tooth, occupies the middle of each row, flanked on either side by several lateral teeth and the marginal teeth along the periphery. The number of teeth per group is expressed by certain formulae. Some radulae, such as the rhipidoglossid type, have large numbers of teeth. The number of teeth may be considerably reduced.

The most important radula types in the Recent prosobranchs are as follows:

rhipidoglossid
the type with the largest number of teeth. Tooth formula approximately 100 : 5 : 1 : 5 : 100
docoglossid
in the Patellacea. Tooth formula 3 : 1 : 1 : 1 : 3
taenioglossid
most mesogastropods. Tooth formula 2–5 : 1 : 1 : 1 : 2–5
stenoglossid (rhachiglossid)
most neogastropods. Tooth formula 1 : 1 : 1 (unknown if marginal or lateral teeth are present)
toxoglossid
median plate reduced, lateral plates stiletto-like, sometimes with barbs

Classification

The neozoological classification of the gastropods is based on the following: (1) torsion and its consequences for the organism (shape of the nervous system); (2) structure of the radula. Neither of these characteristics can be observed in fossils. The palaeontological classification is thus inevitably based on the characteristics of the shell (Fig. 64) and our knowledge of present-day representatives. The most important features are therefore the relation between the peristome and the mantle, which can be deduced from the scars in the shell, and the shape of the peristome. Occasionally scars left by the adductor muscle can be useful.

Fig. 64. Various gastropod shell shapes: (a) convolute (*Acteonella*, × 2); (b) sinistrally coiled (*Physa*, × 0.8); (c) fusiform (*Fasciolaria*, × 2); (d) turretted (*Nerinella*, × 1.1); (e) turretted (*Nerinea*, × 0.5); (f) trochiform (*Calliostoma*, × 2); (g) discoidal, sinistrally coiled (*Heliosoma*, × 1.1); (h) involute (*Bulla*, × 1.2); (i) isostrophic (*Bellerophon*, × 0.7); (j) trochiform (*Turbo*, × 0.5).

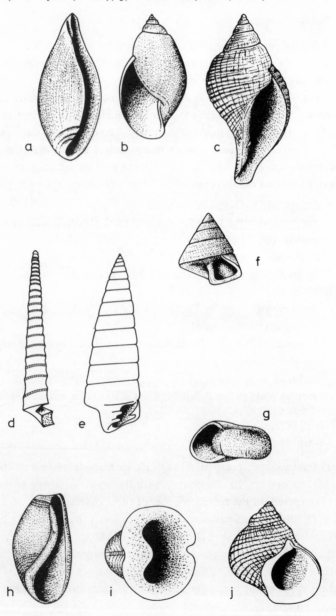

1.   Subclass: Prosobranchia (Streptoneura)
Gastropods with crossed nerve tracts and anterior mantle cavity.
Predominantly marine forms with well developed shells.

1.   Order: Archaeogastropoda
Predominantly herbivorous prosobranchs often with holostomatous
aperture and rhipidoglossid radula. Primitive forms still have two gills and two
kidneys. This order includes the Palaeozoic prosobranchs and their direct
descendants.

1.   Suborder: Bellerophontina
This is an early and rather separate group. The planispiral (isostrophic)
shell has a thick wall and a more or less pronounced slit-band; no known
operculum. Some authors claim that this group did not undergo torsion and
that it belonged to the monoplacophoran lineage.
    Examples:

Bellerophon (Fig. 65, 1): Large with a thick shell; narrow slit-band,
pronounced transverse sculpturing. Occurrence: Ordovician–Lower Triassic.

Euphemites (Fig. 65, 2): Similar to *Bellerophon*, but with wide slit-
band and spiral sculpturing. Occurrence: Upper Devonian–Permian.

Knightites (Fig. 65, 3): Large and thick-shelled; two tube- or lobe-
shaped protrusions on the dorsal peristome in adults. The tubes are thought
to be inhalant or exhalant channels, perhaps corresponding to the siphon of
higher prosobranchs. Occurrence: Devonian–Permian.

2.   Suborder: Macluritina
These are also large and thick-shelled, but have no slit-band. The
whorls are flattened, and the umbilicus is wide open.

    Examples:
Maclurites (Fig. 65, 4): Usually with ledge along the outer periphery
probably corresponding to an exhalant channel. Thick heavy operculum, coiled
anticlockwise, which means that the shells were probably hyperstrophic, i.e.
with the tip pointing downwards. Occurrence: Ordovician–Devonian.

Straparollus: Coiled a little higher with a narrow ridge on the upper
surface. Probably also hyperstrophic and derived from *Maclurites.*

Fig. 65. Order Archaeogastropoda: (1) *Bellerophon* (× 0.8);
(2) *Euphemites* (× 1.7); (3) *Knightites* (× 0.9); (4) *Maclurites* (× 0.4);
(5) *Euomphalus* (× 0.6); (6) *Pleurotomaria* (× 0.4).

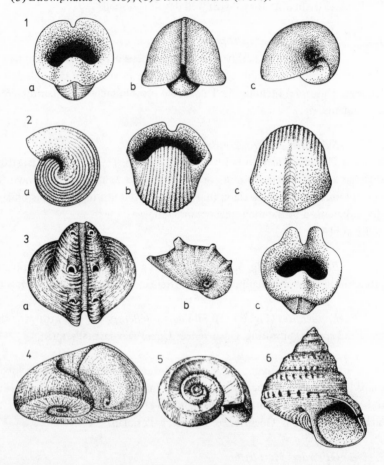

Euomphalus (Fig. 65, 5) is similar but more flattened. Occurrence: Ordovician-Cretaceous.

The other archaeogastropods are mostly coiled into helical shapes. The gills and kidneys are paired. A slit-band is frequently present. Occurrence: Cambrian-Recent.

Examples:

Pleurotomaria (Fig. 65, 6): This could be the ancestral form of all Recent snails. Trochiform shell with fairly wide whorls and a slightly flattened base. The slit-band is usually close to the suture.

The genus *Pleurotomaria* existed in the Jurassic and Cretaceous. Similar Palaeozoic forms are assigned to other genera (*Raphistoma*: Ordovician–Silurian; *Worthenia*: Carboniferous–Triassic).

Murchisonia (Fig. 66, 1): More or less narrow conical shape, numerous whorls the oldest of which are often filled with calcareous secretions. No mother-of-pearl layer; slit-band in the middle of the whorls or at the edge. Aperture sometimes extended downwards to form a 'spout'. Occurrence: Ordovician–Upper Triassic.

Porcellia (Fig. 66, 2): Pseudosymmetrical shell with flat external groove; wide umbilicus; with ribs curved backwards slightly; aperture not widened; weak slit-band, sometimes interrupted. (Moulds of the last part of the last whorl show a keel.) Occurrence: Devonian–Carboniferous.

The status of this genus is still disputed; it may alternatively belong to the bellerophontids.

Kokenella: Similar to *Porcellia*; occurs in the Middle and Upper Alpine Triassic. The cross-section of the whorls is usually angular. Lateral ribs are often thickened into nodes towards the dorsal side; the growth lines behind the mouth border are strongly curved backwards.

Haliotis (Fig. 66, 3): *Haliotis* is directly derived from *Pleurotomaria*, which may not be justified. The final whorl widens out into an ear shape (ear shell). The row of holes (tremata), through which tactile organs from the mantle emanate, corresponds to the slit-band. Mother-of-pearl present on the inside. Occurrence: Cretaceous–Recent, but main development in the Upper Tertiary.

Patella (Fig. 66, 4): *Patella* is regarded as the main representative of the limpets, which have existed since the Silurian. Conical, cap-shaped, usually with more or less pronounced radial ribs; no slit or perforation. Inside, a horseshoe-shaped muscle scar opening to the front. The apex points forwards in some species, and backwards in others. The shell is calcitic on the outside and aragonitic on the inside, but not like mother-of-pearl. Limpets live in shallow water attached to rocks. Occurrence: Cretaceous–Recent.

Trochonema (Fig. 66, 5): Fairly large, conical and fairly elevated. Whorls narrow, descending obliquely at the top as far as the keel and then vertically. Growth lines oblique at the top and straight at the side. Final whorl with two keels, concave between the two keels; below, a third keel surrounding

Fig. 66. Order Archaeogastropoda: (1) *Murchisonia* (× 2); (2) *Porcellia* (× 1); (3) *Haliotis* (× 0.5); (4) *Patella* (× 0.5); (5) *Trochonema* (× 0.7); (6) *Platyceras* (× 0.5); (7) *Subulites* (× 1); (8) *Trochus* (× 0.5); (9) *Nerita* (× 0.4); (10) *Fissurella* (× 1).

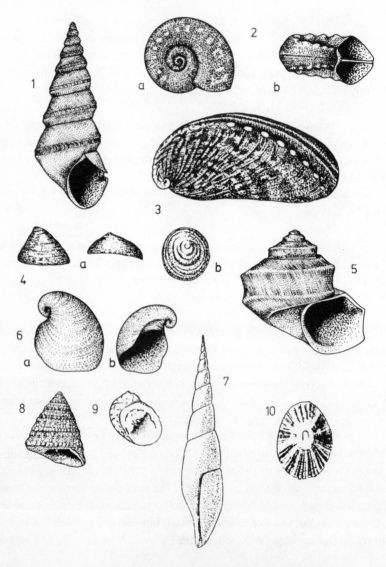

the wide umbilicus. Umbilicus and underside with very oblique growth lines. Holostomatous aperture. Occurrence: Ordovician–Devonian.

Trochonema is regarded as a representative of a now completely extinct group derived from *Pleurotomaria* which was widely distributed in the Palaeozoic, but which had neither a slit nor a slit-band.

Platyceras (Fig. 66, 6): *Platyceras* is a representative of those forms with secondary evolute uncoiling. The shell is often thin, calcitic and quite large, tall hat- to horn-shaped and more or less irregular. The initial whorls are coiled normally, while the final whorls widen out rapidly. Sometimes the shells are furnished with hollow spines.

Some of the platyceratids had a coprophagous, pseudoparasitic lifestyle near the anal openings of crinoids and cystoids and lived off their excrement; generally sedentary lifestyle. Occurrence: Silurian–Permian.

The subgenus *Platyceras (Orthonychia)* was completely straight and also lived at the anal openings of crinoids. Occurrence: Silurian–Permian.

The following two orders are representatives of the Trochacea, which have existed since the Triassic and probably evolved from the Trochonematacea. Like *Trochonema* they have no slit-band. Their aperture is holostomatous. They are regarded as possible ancestral forms at the root of the neogastropods.

Trochus (Fig. 66, 8): Base flattened, angular; aperture usually triangular to rectangular. Mouth borders not connected. Umbilicus partly (in the upper part) or completely filled by a callous coating (false umbilicus); spindle (columella) inserted in the false umbilicus, coiled at the top and usually furnished with a tooth at the bottom. Occurrence: Miocene–Recent.

Fissurella (Fig. 66, 10): Shell conical, porcellaneous; with exhalant perforation at apex; horseshoe-shaped muscle scar, open anteriorly. Occurrence: Oligocene-Recent.

Subulites (Fig. 66, 7): Slender, spindle-shaped. Whorls almost smooth; aperture narrow, elongate, with small notch. Occurrence: Ordovician–Devonian.

This genus represents an isolated group which suddenly appeared in the Ordovician and perhaps already belongs to the caenogastropods. With the younger pseudomelanids this group extended right into the Tertiary.

Naticopsis: Rounded to oblique egg shape. Short spiral with few whorls; the inner walls of the whorls are not resorbed. The final whorl is very large, the aperture oblique; the mouth edges are connected by a protruding inductura (smooth shell extending from the inner side of the aperture). Occurrence: Devonian-Triassic.

Nerita (Fig. 66, 9): The inner walls of the whorls are resorbed, other-wise *Nerita* is similar to *Naticopsis*. Usually strongly sculpted; inhabits the rocks of tidal zones. Occurrence: Upper Cretaceous-Recent.

Neritaceans (*Naticopsis, Nerita*) stand apart from other archaeogastropods.

2.   Order: Caenogastropoda

Prosobranch gastropods with one gill, one kidney and often with a siphonal canal. This order includes the majority of the Mesozoic and Cenozoic prosobranchs, but few Palaeozoic prosobranchs.

1.   Suborder: Mesogastropoda

Aperture mostly with siphonal canal. Usually with taenioglossid radula with tooth formula 2-5:1:1:1:2-5. Oldest, though very doubtful, representatives from the Ordovician.

Examples:

Loxonema (Fig. 67, 1): Slender and turretted with numerous more or less arched whorls. Simple aperture the bottom of which protrudes forward. S-shaped growth lines. Occurrence: Ordovician–Carboniferous.

The forms based on *Loxonema* are regarded as an important starting point for the caenogastropods. They probably originated from the murchisoniids by reducing the slit-band. The latter already had this tendency.

Viviparus (Fig. 67, 2): Shell conical to spherical in shape, narrow or covered umbilicus, whorls usually smooth in the Recent forms. The closely related fossil representatives (*Tulotoma*), particularly in the Pliocene of Europe and Asia Minor, also have simple or nodular keels. Worldwide inhabitors of fresh waters. Numerous species have perhaps existed since the Carboniferous, but most certainly since the Jurassic.

Littorina (Fig. 67, 5): Some species of *Littorina* (periwinkle) are viviparous. Shell usually thick-walled, without umbilicus, conical with blunt apex. Usually with spiral stripes; aperture oblique. Sturdy aperture, not continuous, sharp periphery, top protruding forwards; edge of the spindle shiny and folded back across the umbilicus. Inhabits the littoral and tidal zones and feeds on algae. Occurrence: Palaeocene–Recent.

Hydrobia (Fig. 67, 3): Small, elongate and conical. Simple peristome, not thickened, continuous. *Hydrobia* predominantly lives in fresh-water, occasionally in brackish-water or even in marine habitats. One species lives on damp rocks. Occasionally occurs as a rock-builder (*Hydrobia* strata in the Mainz basin). Occurrence: Jurassic–Recent.

The similar Rissoidae are predominantly marine.

Fig. 67. Suborder Mesogastropoda: (1) *Loxonema* (×0.33); (2) *Viviparus* (×0.7); (3) *Hydrobia* (×0.5); (4) *Turritella* (×1); (5) *Littorina* (×1); (6) *Vermetus* (×1); (7) *Cerithium* (×1.5); (8) *Nerinea* (×0.7).

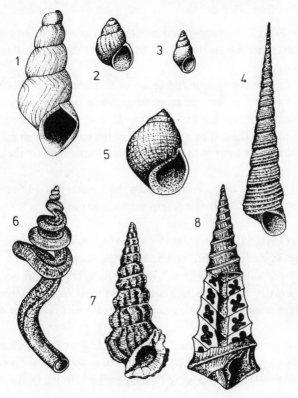

Turritella (Fig. 67, 4): Tall turretted shell with 8–20 whorls and spiral sculpturing. Holostomatous aperture, not oblique ('turret shell'). Occurrence: Oligocene–Recent (Family Turritellidae Devonian–Recent).

Vermetus (Fig. 67, 6): *Vermetus* is an aberrant sessile form with a short spiral part to begin the shell and an irregular final whorl. It builds gastropod reefs. Occurrence: Tertiary–Recent, perhaps even Triassic–Recent.

Tenagodes (= *Siliquaria*): Similar to *Vermetus*, but the initial whorls are also irregular. A slit which is partly closed extends along the whole length. *Tenagodes* has a parasitic lifestyle in sponges. Occurrence: Triassic–Recent.

Cerithium (Fig. 67, 7): Similar to *Turritella*, but the aperture has a notch and is set obliquely, oval; transverse sculpturing with nodes or spines.

Occurrence: rare in the Upper Cretaceous; Tertiary-Recent. The closely related giant form *Campanile* from the French Middle Eocene can reach a length of 0.5 m.

Nerinea (Fig. 67, 8): Large, similar to *Cerithium*, but sculpturing less pronounced, wrinkles present on the spindle and walls of the whorls. Whorls sometimes concave.

The nerineids lived in the Jurassic and Cretaceous. Their taxonomic status is still uncertain. Some forms were 20 times taller than wide and thus extremely turretted. The external surface has little or no sculpturing. Characterised by spiral folds along the spindle and along the walls of the whorls, which may be so pronounced that they almost obliterate the cavity (particularly in the subgenus *Itieria*).

Twenty-one genera have been described from the Jurassic and 25 from the Cretaceous, but neither ancestors nor descendants are known.

Hipponix (Fig. 68, 1): Shell bowl- or cap-shaped. Apex approximating the posterior edge; horseshoe-shaped muscle scar wide open to the front. The animal lives attached to stones, snail and bivalve shells, corals, calcareous algae, etc., and creates a cavity in the substrate and anchors itself via its adhesive muscles which leave behind impressions. Occurrence: Upper Cretaceous-Recent.

Atlanta (Fig. 68, 2): Shell small, dextrally coiled, keeled, tendency towards bilateral symmetry. Pelagic form, widespread in warm seas. *Atlanta* is a representative of the 'Heteropoda' described by earlier authors. Occurrence: Upper Cretaceous-Recent.

Crepidula (Fig. 68, 4): Lives attached to rocks, bivalve shells, etc., which often influences its shape. An internal septum covers the rear part of the aperture. Occurrence: Upper Jurassic-Recent.

The following four genera are members of the Strombacea. These usually have radial sculpturing and a long spindle with a gutter-shaped protruding aperture and a pronounced canal. The peristome is usually folded back and extended into wing-like projections with finger-like processes. The Strombacea first appeared in the Triassic.

Xenophora (Fig. 68, 5): Conical shape, peristome very oblique. Characterised by the agglutinous inclusion of foreign bodies in the deeply sunken sutures. Occurrence: Upper Cretaceous-Recent.

Fig. 68. Suborder Mesogastropoda: (1) *Hipponix* (×1); (2) *Atlanta* (×2); (3) *Bursa* (×1); (4) *Crepidula* (×0.5); (5) *Xenophora* (×0.5); (6) *Cypraea* (×0.5); (7) *Aporrhais* (×0.5); (8) *Natica* (×1); (9) *Strombus* (×0.5); (10) *Galeodea* (×0.8); (11) *Tonna* (×0.5).

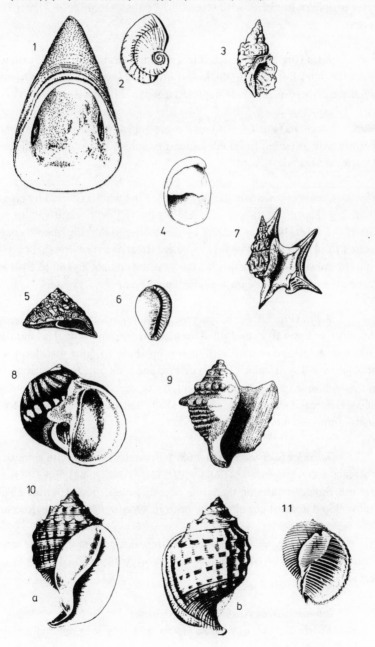

Aporrhais ('pelican's foot shell') (Fig. 68, 7): Shell conical and turretted; aperture furnished with finger-like projection at the top which may be free or lie close up against the whorls; at the bottom commonly a fairly long projection. Outer periphery furnished with several projections. Occurrence: Upper Jurassic-Recent.

Bursa (Fig. 68, 3): Shell thick, spire moderately high; aperture with thickened outer lip; inner lip thickened and curved; with two opposite varices (thickenings) across the whole shell. Occurrence: Eocene-Recent.

Strombus (Fig. 68, 9): Large conical thick-walled shell, elongated aperture with a straight lower edge. Outer periphery extended and thickened. Occurrence: Eocene-Recent.

Cypraea (cowrie shell) (Fig. 68, 6): First whorls covered by large final whorl, egg-shaped, aperture narrow and slit-shaped, denticulate, spindle teeth. Part of the outer shell is enveloped by the mantle so that the smooth coating forms a third shell layer. The first representatives of the cowrie shells are known from the Jurassic. They are divided into (far too) numerous (up to 140) genera. Occurrence of the genus *Cypraea*: Miocene-Recent.

Natica (Fig. 68, 8): Bulging shell with holostomatous aperture, spindle edge more or less warty and folded back across the umbilicus. The animals burrow through sand or mud with their propodium covering their head, searching for bivalves and snails which they bore into and suck out. A disc-shaped gland on the underside of the propodium secretes an acid calcium-carbonate-dissolving fluid which eats a circular hole into the shell. Occurrence: Palaeocene-Recent. Similar forms are much older.

Galeodea (Cassidaria) (Fig. 68, 10): Arched whorls with smooth or tubercular spiral stripes, sometimes with a varix (thickening) opposite the aperture. Small wrinkles on the inside edge of the aperture. Aperture channel almost closed and not cut out at the bottom. Occurrence: Eocene-Recent.

Tonna (Fig. 68, 11): Shell large, thin, ovate; spire short; last whorl very large, longitudinally striate; aperture with short, wide siphonal canal. Occurrence: Tertiary-Recent.

2.   *Suborder: Neogastropoda (Stenoglossa)*
Aperture siphonostomous, mostly with very long siphonal canal; middle Ordovician-present day.

Murex (Fig. 69, 1): Shell rounded to egg-shaped and fairly tall. Whorls tuberous because of the presence of numerous spiral stripes and growth lines; numerous tuberous or spiny thickenings. Aperture egg-shaped with narrow, elongated, almost completely closed canal at the bottom. Predatory lifestyle. Occurrence: Miocene–Recent. This form converges to the Tonnaceae but has a different radula.

Buccinum ('whelk') (Fig. 69, 3): Final whorl large and bulging, rounded at the bottom, without an umbilicus. Wide aperture with very short canal cut out at the end. Occurrence: Oligocene–Recent.

Fusinus (Fig. 69, 2): Slender, spindle-shaped shell, ranging from small to very large in size. Final whorl with a more or less elongated neck with an open canal. Outer edge sharp with small folds on the inside. Occurrence: Upper Cretaceous–Recent.

Fig. 69. Suborder Neogastropoda: (1) *Murex* (×0.7); (2) *Fusinus* (×1); (3) *Buccinum* (×0.5); (4) *Tudicla* (×0.5); (5) *Turris* (×1); (6) *Conus* (×0.5). Superorder Opisthobranchia: (7) *Acteonina* (×0.5); (8) *Acteonella* (×0.5).

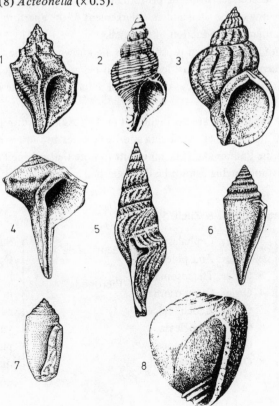

Tudicla (Fig. 69, 4): Bulging shell, aperture with straight and very long canal, inner lip with one fold. Occurrence: Upper Cretaceous–Recent.

Turris (= *Pleurotoma*) (Fig. 69, 5): Spindle-shaped shell; outer edge of the aperture with more or less deep indentation (anal sinus) at the top. The growth lines follow the pattern of the anal sinus and thus create a more or less prominent anal band. No spindle folds. Toxoglossid radula. Occurrence: Cretaceous–Recent.

Conus (Fig. 69, 6): Conical shell; internal wall of the whorls always resorbed; short coiled section, long narrow aperture with notch. Toxoglossid radula. Rather poisonous. Occurrence: Cretaceous–Recent.

2.   Subclass: Euthyneura
In the members of this subclass the phenomenon of chiastoneury has partly been reversed during ontogeny by the backward (clockwise) rotation of the visceral sac. The respiratory organs, one gill or the lungs, are again on the right-hand side behind the heart; the animals are thus secondarily orthoneural. They are probably descended from the prosobranchs. The Euthyneura are characterised by a shortening of the shell and an enlargement of the aperture as well as a tendency to progressive reduction of the shell.

Forms which are reminiscent of prosobranchs are known since the Carboniferous. Of the Recent gastropods just less than half belong to this subclass. Because of the widespread tendency towards shell reduction, only the Pulmonata play a role as fossils.

In general, two superorders are distinguished in the fossil record: representatives of the purely marine Opisthobranchia and fresh-water and terrestrial representatives of the Pulmonata. It is important to note that marine Pulmonata and primitive Opisthobranchia cannot be distinguished as fossils.

Classification (Wenz & Zilch, 1959/60)

'Tectibranchia' with shells	1. Order: Cephalaspidea 2. Order: Anaspidea	'Pleurocoela'	Superorder Opistho-branchia in the narrower sense; purely marine
	3. Order: Thecosomata 4. Order: Gymnosomata	'Pteropoda'	
	5. Order: Acochlidiacea		
	6. Order: Sacoglossa		
Acoela	7. Order: Notaspidea 8. Order: Nudibranchia		

| Pulmonata | { 9. Order: Basommatophora | Superorder |
| | { 10. Order: Stylommatophora | Pulmonata |

1.   *Superorder: Opisthobranchia*

The most primitive opisthobranchs are the Cephalaspidea, with forms such as *Acteonina* (since the Carboniferous) with a massive shell and an operculum. Some *Sacoglossa* forms have two-part bivalve-like shells.

Shell reduction has been carried furthest by the Anaspidea and Nudibranchia, some of which have no shell at all.

The group of deep-sea snails formerly known as pteropods, in which the foot has been modified into a pair of fins and whose shells are sometimes enriched as pteropod muds, are now divided into two orders. The term Pteropoda unites the two orders Gymnosomata (without shells) and Thecosomata (often with shells). They are small, free-swimming pelagic opisthobranchs in which the foot has been transformed into a pair of wing-like fins. They frequently occur in large shoals and form an important part of the diet of large marine vertebrates. Their shells are mostly thin and translucent, conical to planispiral in shape, and may accumulate as pteropod ooze on the ocean floor at depths between 1000 and 2700 m.

Acteonina (Fig. 69, 7): More or less conical shell, spindle with small folds disappearing towards the outside. Short, stepped coiled section. Occurrence: Carboniferous–Jurassic.

Acteonella (Fig. 69, 8): Fairly large with thick shell; large final whorl, smooth with three horizontal spindle folds. Occurrence: Triassic–Upper Cretaceous.

2.   *Superorder: Pulmonata*

7000 Recent, 800 fossil species, 1000 genera, 50 families; very diverse, derived from the opisthobranchs (or vice versa). Known since the Carboniferous, abundant since the Tertiary, in their prime at the present time. Stratigraphically important in the Tertiary.

1.   *Order: Basommatophora*

Eyes not on stalks; live predominantly near or in water.

Planorbis ('ramshorn snail') (Fig. 70, 1): Numerous whorls, low conical shape to virtually planispiral. Occurrence: Jurassic–Recent, abundant in the Tertiary.

Fig. 70. Superorder Pulmonata: (1) *Planorbis* (×1.5); (2) *Lymnaea* (×1); (3) *Clausilia* (×2); (4) *Physa* (×0.5); (5) *Pupilla* (×6); (6) *Ancylus* (×1.5); (7) *Helix* (×1).

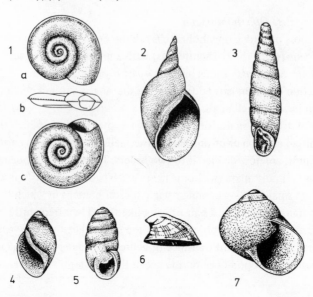

Lymnaea (Fig. 70, 2): Large final whorl, wide aperture; thin-shelled; dextrally coiled. Occurrence: Upper Jurassic–Recent.

Physa (Fig. 70, 4): Like *Lymnaea*, but sinistrally coiled. Occurrence: Upper Jurassic–Recent.

Ancylus (Fig. 70, 6): Small cap-shaped shell; apex moved backwards; radial stripes on the upper side; wide egg-shaped aperture. *A. fluviatilis* is a guide fossil for the Boreal fresh-water *Ancylus* Sea (Baltic Sea around 6000 B.C.). Occurrence: Tertiary–Recent.

2.   Order: *Stylommatophora*
        Stalked eyes; terrestrial habitat.

Helix (Fig. 70, 7): Large, spherical, with covered umbilicus; ancestral form (*Archaeozonites*) known since the Carboniferous, but genus *Helix* known only since the Cretaceous. *Helix pomatia*: present-day edible snail or Roman snail.

Pupilla (Fig. 70, 5): Shaped like a bee-hive; 2.5–10 mm tall with aperture folds. Occurrence: Eocene–Recent.

Clausilia (Fig. 70, 3): Slender, conical, small, usually sinistrally coiled. Aperture narrowed by folds. Calcareous lid (clausilium). Occurrence: Eocene– Recent.

3.   Class: Scaphopoda (tusk-shells)
        Scaphopods are molluscs of very simple organisation with a heart, but without eyes or gills. They have a short radula with five teeth per transverse row. The foot is designed for burrowing. Mantle and shell have fused ventrally to form a tube. The burrowing foot is reminiscent of the lamellibranchs (Fig. 71).

Fig. 71. Organisation of a scaphopod.

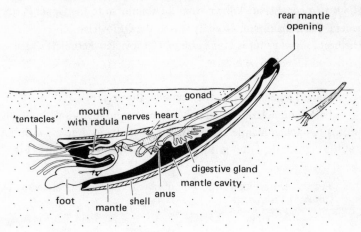

Scaphopods burrow in soft sediments in a slightly oblique position so that the thin upper end protrudes into free water. They eat small organisms such as foraminifera in the substrate, which are caught by flexible tentacle-like threads (captacula). Respiration is effected by the whole body surface, with water being periodically sucked into and expelled from the end of the tube. Scaphopods live only in the sea in the neritic to bathyal zones, rarely in the littoral zone.
        Occurrence: Ordovician (?), Devonian–Recent; of greater importance in the Tertiary.
        Two families are distinguished:

1.   Family: Dentaliidae
        The shell tapers evenly like an elephant's tusk; the widest diameter is at the aperture. Shells can measure up to 12 cm. *Dentalium* is the main genus with several subgenera.

2.   *Family: Siphonodentaliidae*

Members of this family are considerably smaller, measuring 1-2 cm at most. Their shells taper towards the aperture. They inhabit deeper regions of the sea. *Siphonodentalium.*

4.   Class: Rostroconchia

An extinct class of molluscs characterised by a bivalve adult shell without a ligament. The two valves are connected across the dorsum by one or more cõntinuous shell layers. There is only one centre of calcification, situated in the univalve protoconch. The shell becomes bivalved by the excessive growth of the lateral lobes of the mantle and shell.

The Rostroconchia are known from the Cambrian to the Upper Permian with 34 genera. Their maximum diversity was in the Ordovician.

The best known genus is *Conocardium* (Ordovician-Permian) which was formerly considered to be bivalve.

5.   Class: Lamellibranchia (Pelecypoda, Bivalvia)

Lamellibranchs are bilaterally symmetrical molluscs without a head, but with a two-lobed mantle and a two-part shell. In accordance with the orientation of the animal, the latter is divided into a right and a left valve (Fig. 72). On

Fig. 72. Dorsal view of a lamellibranch. A, area; Li, ligament; Lu, lunule; U, umbo.

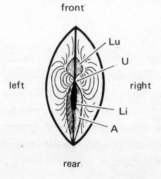

the dorsal side lies the hinge, a variable system of teeth and sockets which holds the valves together. The dorsal ligament serves to open the valves, while closure is brought about by muscle contraction. The paired gills are situated beneath the mantle lobes (Fig. 73). The siphon, which in many bivalves represents a double tube originating from the fused mantle edges and serves as both an incurrent and an excurrent channel, is directed backwards. The mantle is attached to the shell

Fig. 73. Transverse section of a lamellibranch. F, foot; G, gills; H, heart; Hy, hypostracum; L, ligament; M, mantle; Pe, periostracum; Pr, prismatic layer.

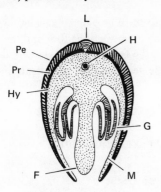

along the pallial line, which may be entire (integripalliate) or interrupted by an indentation towards the rear of the shell (sinupalliate) if there is a well developed siphon (Fig. 74).

The muscular foot, which lies ventrally, can be used for crawling, leaping and boring. At the ventral end there is often a byssus gland which secretes very tensile byssus threads used by the bivalve to anchor itself to a solid substrate.

Two adductor muscles which extend from one valve to the other and close the shell are usually present. Bivalves with two adductor muscles of about the same size are known as homomyarian (= dimyarian, isomyarian); those where only the rear adductor muscle is well developed are called anisomyarian (Fig. 74).

Fig. 74. Diagrammatic representation of the adductor muscle scars and the shape of the pallial line in lamellibranchs.

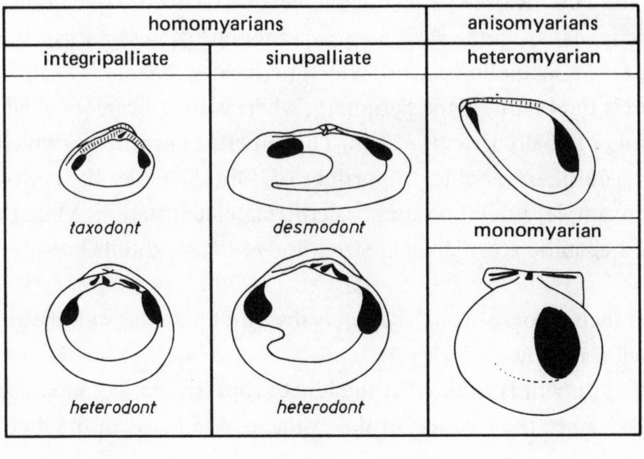

homomyarians		anisomyarians
integripalliate	sinupalliate	heteromyarian
taxodont	*desmodont*	monomyarian
heterodont	*heterodont*	

Of the latter forms those with an atrophied front adductor muscle are called heteromyarian, those with only one adductor muscle (the front one being completely absent) monomyarian.

The lamellibranch nervous system consists of three pairs of ganglia: head, pedal and visceral ganglia. The gills are situated between the foot and the mantle.

According to their degree of complexity, the following gill types are distinguished, some of which form the basis of the neozoological classification of the lamellibranchs (Fig. 75):

1. protobranch (Cryptodonta);
2. filibranch (many Taxodonta and Dysodonta);
3. eulamellibranch (Heterodonta and most Desmodonta);
4. septibranch (Poromyacea).

Fig. 75. Schematic representation of lamellibranch gill types.

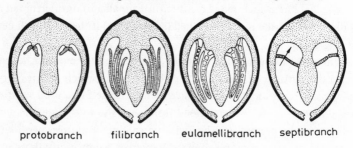

protobranch   filibranch   eulamellibranch   septibranch

The shell

The shell consists of calcium carbonate and conchiolin with varying amounts of magnesium, phosphorus and silica. The formation of the shell begins with the thin, flexible periostracum which consists of conchiolin. Sometimes this alone makes up the shell material (anodontids in salt seas). It is connected to the conchiolin network which permeates the whole shell. The periostracum is thickest near the periphery, whereas the calcareous shell material becomes thinner radially from the beak. The mineral constituents, which may be calcitic or aragonitic or variable proportions of both, lie under the periostracum. Oysters, for example, exhibit a purely calcitic platelet structure, while *Cardium* has a purely aragonitic cross-lamellar structure. *Mytilus* exhibits both types of structure.

The larval shell (prodissoconch) is equivalve and bilaterally symmetrical; the adult shell is mostly equivalve.

The elastic ligament is situated at the hinge (top). In the Taxodonta successive growth stages lead to the creation of the ligament area between the umbones.

The shell is usually prosogyral (umbo curved so that the beak points anteriorly), occasionally opisthogyral (umbo curved so that the beak points posteriorly: e.g. in *Nucula, Trigonia, Donax*).

The left and right valves are recognised by looking at the shell from above with the dorsal side upwards and the anterior end pointing away from the observer (Fig. 72). The following features serve to identify the posterior:

1. the ligament is usually only developed at the rear end;
2. the lunule, an elongated oval area on both sides of the hinge with variable sculpturing, lies in front of the umbo;
3. the rear muscle scar is frequently more pronounced than the front one, which may be altogether absent;
4. the mantle sinus is at the rear;
5. the rear portion of the shell is usually more developed;
6. the umbones are usually curved towards the front.

The ligament, actually a special element of the shell, can be external or internal (in the latter case it lies in one or several special pits) and consists of a harder outer rind (actual ligament or lamellar layer) and a very elastic inner layer (resilium or fibrous layer). The tendency of the resilium to retain its shape makes the shells gape when the adductor muscles relax. If the ligament is accommodated in special pits (nymphs) on the inside, it is made up only of the resilium. The outer ligament is either amphidetic (in front of and behind the umbo) or opisthodetic (only behind the umbo).

The hinge prevents the two valves from moving with respect to each other. The most important types of hinge are (Fig. 76):

1. *Taxodont*: numerous subequal parallel teeth, oblique or vertical with respect to the edge of the hinge, fit into sockets in the opposing valve.
   (a) Ctenodont, pseudoctenodont (teeth converging towards the inside);
   (b) actinodont (teeth converging towards the outside).
2. *Dysodont*: very small weak teeth or rather fine fluting of the periphery.
3. *Heterodont*: in each valve, a small number of different-shaped teeth and the corresponding sockets. The hinge, or cardinal, teeth lie under the umbones, the lateral teeth laterally more or less in parallel with the hinge line.
4. *Preheterodont (diagenodont)*: a few ridges converging towards the umbo (*Lyrodesma*).
5. *Isodont*: two teeth and two sockets, respectively, symmetrically placed either side of the ligament pit (*Spondylus*).
6. *Schizodont*: a triangular hinge tooth of the left valve fits into a socket

Fig. 76. Lamellibranch hinge types: (a) taxodont (actinodont; *Lyrodesma*, × 2); (b) schizodont (*Trigonia*, × 1); (c) dysodont (*Mytilus*, × 0.3); (d) taxodont (ctenodont; *Nucula*, × 1.5); (e) pachydont (*Diceras*, × 0.5); (f) isodont (*Spondylus*, × 1); (g) heterodont (cyrenoid; *Corbicula* (= *Cyrena*), × 0.6); (h) heterodont (lucinoid; *Lucina*, × 0.4); (i) pachydont (*Hippurites*, × 0.3); (j) desmodont (*Mya*, × 0.9).

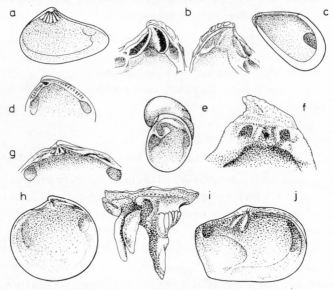

between two usually divergent and sometimes fluted teeth of the right valve (*Trigonia*).

7. *Desmodont*: hinge teeth absent; the ligament is sometimes carried by inwardly projecting processes (chondrophores).

8. *Pachydont*: one to three unsymmetrical peg-shaped projections fit into corresponding sockets in the opposing valve (rudistids).

The primary function of the shell is protection, which is particularly important in view of the predominantly sessile lifestyle. The shape of the shell is dependent on the lifestyle as well as on the phylogenetic status of the animal. There are numerous variations within an evolutionary line, so that homeomorphisms are common. With a few exceptions, the evolutionary lines were fixed as early as the Palaeozoic.

The most important ecological types are:

1. Epifauna: mobile, sometimes capable of swimming; one valve fixed or attached by means of byssus threads.

2. Infauna: burrowing bivalves with more or less long siphons; boring bivalves.

Feeding

Most bivalves are filter-feeders and live off suspended phytoplankton (diatoms, dinoflagellates, etc.) which is trapped by the gills. Sediment-feeders utilise the organic material present in the sediment, such as bottom-dwelling diatoms. Only the Poromyacea (Septibranchia) are carnivores; they feed on small worms and crabs.

Lifespan

Information about the lifespan of bivalves has only been available for a few years. For example:

> *Sphaerium*   <1 year
> *Tellina*   5 years
> *Mytilus edulis*   8–10 years
> *Ostrea*   >12 years
> *Cardium edule*   14 years
> Large fresh-water bivalves   70–100 years
> *Tridacna*   100, 200, 300 years?

Table 5. *Summary of the most important diagnostic features of the lamellibranchs*

Subclass	Gill type	Hinge	Adductor muscle	Pallial line
Palaeotaxodonta	Protobranch	Taxodont	Homomyarian	?
Cryptodonta	Protobranch	Toothless or with indentations	Homomyarian or heteromyarian	Integri- or sinupalliate
Pteriomorphia	Filibranch	Taxodont, isodont or ± toothless	Homomyarian to monomyarian	Integripalliate
Palaeo-heterodonta (= Schizo-donta)	Eulamellibranch	Heterodont	Homomyarian	Integripalliate
Heterodonta	Eulamellibranch	Heterodont (Veneroida), pachydont (Hippuritoida)	Homomyarian	Integripalliate or sinupalliate
Anomalodesmata	Eulamellibranch or septibranch	Often toothless (chondrophore)	± Homomyarian	Integripalliate

Phylogeny

The oldest identifiable fossil bivalve seems to be *Fordilla troyensis* Barrande, well known from the Lower Cambrian rocks of New York State, but apparently widespread in the Lower Cambrian. The small shells (5 mm long) have adductor and pedal retractor muscle scars and an indented pallial line. *Fordilla* may have given rise to the Actinodonta (Pojeta *et al.*, 1973).

Lamellodonta simplex Vogel from the Lower Middle Cambrian of Spain is no longer considered to be a bivalve; it turned out to be a brachiopod (*Trematobolus simplex* (Vogel): Havlíček & Kříž, 1978).

The most important groups appeared during the course of the Ordovician Period. The Palaeoheterodonta occupied a central position in the Lower Ordovician, together with the Actinodonta which possibly gave rise to the Palaeotaxodonta (= Nuculida), previously regarded as the most primitive bivalves. As early as the Middle Ordovician the first Pseudoctenodonta appeared with *Parallelodon* and simultaneously with the first Anomalodesmatacea. The first Heterodonta appeared in the Upper Ordovician with *Cypricardinia*.

Cryptodonta are known from the Silurian (*Cardiola*). The basic division is thus complete early in the Upper Palaeozoic. Only the rudistids (Upper Jurassic and Cretaceous) represent a younger line which developed from heterodont ancestral forms.

1.   Subclass: Palaeotaxodonta
The Palaeotaxodonta are bivalves with a simple taxodont (ctenodont) hinge and protobranch gills; they are infaunal and epifaunal detritus-feeders.

Fig. 77. Palaeotaxodonta: (1) *Nucula* (× 3); (2) *Yoldia* (× 1).

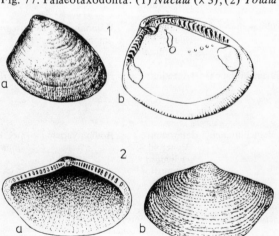

They represent an isolated group without descendants.
Examples:

Ctenodonta: Shell oval to elongated; surface smooth or with fine
concentric lines (*Praectenodonta* has pronounced concentric sculpturing).
Curved row of teeth; teeth bent or straight; integripalliate; ligament external.
Occurrence of the family represented by *Ctenodonta*: Ordovician-Carboniferous.

Nucula (Fig. 77, 1): Oval to triangular, often with truncated posterior
end, but not gaping; surface with concentric ribs, opisthogyral. Inner ligament
on a resilifer. The teeth are slightly curved. Present-day nuculids crawl on
a flattened foot; they occur down to depths of 3000 m. Occurrence: genus
Nucula Upper Cretaceous-Recent; the species-rich Family Nuculidae
Ordovician-Recent.

Nuculana (syn. *Leda*): Similar to *Nucula*, but posterior end produced
with shallow pallial sinus; posterior keel often present. This genus occurs predominantly in Arctic oceans. Occurrence: genus Triassic-Recent; the Family Nuculanidae
Devonian-Recent.

Yoldia (Fig. 77, 2): Similar to *Nucula*, but gaping a little at the rear,
with pallial sinus. '*Yoldia' arctica* (now *Portlandia arctica*) is the guide fossil
for the post-glacial *Yoldia* Sea in the area of the Baltic Sea.

2.   Subclass: Cryptodonta
        The Cryptodonta comprise the so-called Palaeoconcha, small thin-
shelled bivalves with toothless hinges and protobranch gills, known mainly from
the Lower and Middle Palaeozoic.
        Examples:

Cardiola (Fig. 78, 1): Small, quite convex, thin-shelled valves; lattice-
like sculpturing with radiating and concentric grooves; triangular area below
the umbones. *C. cornucopiae* (= *C. interrupta*) is a guide fossil for the Silurian
Period.

Buchiola (Fig. 78, 2): Similar to *Cardiola*, but 'retrostriate' on wide
radiating costae. *B. retrostriata* is a guide fossil for the Upper Devonian.

Solemya: The only Recent representative of the Subclass Cryptodonta.
Thin-shelled, elongated, gaping on both sides. Toothless; inner ligament in

Fig. 78. Cryptodonta: (1) *Cardiola* (×1); (2) *Buchiola* (×6).

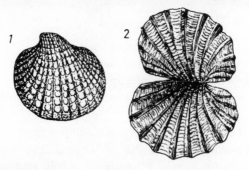

nymphs; section in front of the umbo longer than that behind the umbo. Front adductor muscle scar larger than the rear one. Indistinct pallial line. Burrows deep in the sand. Occurrence: Cretaceous–Recent.

3.   Subclass: Pteriomorphia

This subclass includes the orders Arcoida (= Pseudoctenodonta), Pterioida (= Monomyaria) and Mytiloida (= Anisomyaria). Its members mostly lead an epibenthic life; few bore into hard substrates. Many live attached by means of byssus threads throughout their lives.

Examples of the Order Arcoida (taxodont hinge, but filibranch gills):

Arca (Fig. 79, 1): Large, thick-walled, trapezoid shell with concentric and radial sculpturing; straight hinge line. Very obvious ligament area with bent grooves. Ligament external and amphidetic (present on both sides of the umbo). Pallial line entire (integripalliate); isomyarian. Occurrence: (? Triassic)Jurassic–Recent. These days, the diverse arcaceans mostly live in warm seas.

Glycymeris (syn. *Pectunculus*) (Fig. 79, 3): Thick-shelled, equilateral, round with crenated margin ('hidden ribbing'). Ligament area similar to that of *Arca*. Hinge teeth oblique, converging towards the inside of the valve. Rear muscle scar bounded by a ridge to the front, both adductor muscles roughly equal. The present-day representatives of this genus are abundant in warm seas. Occurrence: Cretaceous–Recent.

Limopsis: Similar to *Glycymeris*, but smaller and less convex; rear adductor muscle scar larger than the front one. Present-day distribution: predominantly deep sea and cold seas. Occurrence: Jurassic–Recent.

Parallelodon (Fig. 79, 2): Thick-walled, strongly convex rectangular shell; umbones displaced well to the front. Shallow ligament area below the

Fig. 79. Pteriomorphia (Order Arcoida): (1) *Arca* (×0.5); (2) *Parallelodon* (×0.7); (3) *Glycymeris* (×1).

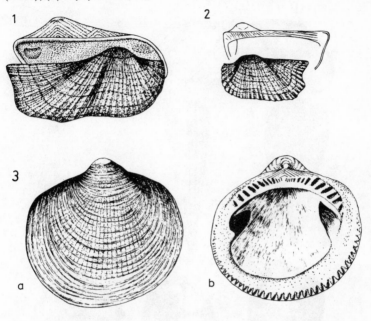

umbo. Hinge with three to seven notches to the front and two or three long lateral teeth. Occurrence: Ordovician–Jurassic.

Examples of the Order Mytiloida (= Anisomyaria) which are characterised by the presence of a weak anterior muscle scar and a more or less toothless hinge line:

Mytilus (common mussel) (Fig. 80, 1): Very inequilateral; the umbo is shifted almost completely to the front; hinge line toothless or slightly indented; long narrow opisthodetic ligament; small anterior adductor scar; attached by byssus threads. Occurrence: Triassic–Recent.

Lithophaga (syn. *Lithodomus*) (Fig. 80, 2): Shell almost cylindrical; umbo shifted well to the front; hinge without teeth; smooth shell surface. Bores into limestones by chemical means (using respiratory carbon dioxide). Occurrence: ? Carboniferous; Tertiary–Recent.

Modiolus: Shell similar to that of *Mytilus* but with blunt and not quite terminal umbo. Front end expanded in a wing-like manner. Toothless hinge. Occurrence: Devonian–Recent.

Fig. 80. Pteriomorphia (Order Mytiloida): (1) *Mytilus* (× 0.5);
(2) *Lithophaga* (× 0.6); (3) *Pinna* (× 0.15). (Order Pterioida): (4) *Eurydesma* (× 0.33).

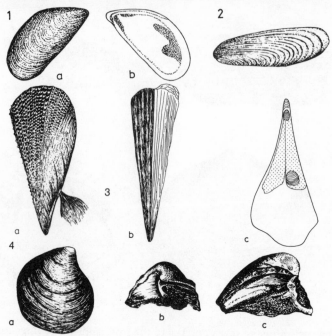

Pinna (fan mussel) (Fig. 80, 3): Shell more or less triangular, gaping widely at the rear; long straight ligament. Each valve consists of two faces which may also be found in isolation. Some forms with a very thick calcitic prismatic layer. In many cases only moulds with a rectangular cross-section are found to be preserved. Fan mussels can grow up to 70 cm in length. They live with their pointed front end buried in the sediment, attached by byssus threads. Occurrence: Carboniferous–Recent.

Trichites: Large, irregular trapezoid shape. Shell very thick and consisting of fibrous calcite; usually preserved only as fragments. Occurrence: Jurassic–Cretaceous.

Examples of the Order Pterioida (Monomyaria) characterised by only one (the posterior) adductor scar:

Eurydesma (Fig. 80, 4): Shell smooth and thick, oval with deep lunule. Index fossil for the marine intercalations of the Lower Permian of the continent of Gondwana.

Fig. 81. Pteriomorphia (Order Pterioida): (1) *Pterinea* (× 0.7); (2) *Pteria* (a: *Pteria hirundo*, Recent, × 0.33; b, c: *Pteria (Rhaetavicula) contorta*, × 1); (3) *Gervillia (Hoernesia) socialis* (× 0.7); (4) *Posidonia* (× 0.7); (5) *Inoceramus* (× 0.33); (6) *Isognomon* (× 0.5); (7) *Daonella* (× 0.6).

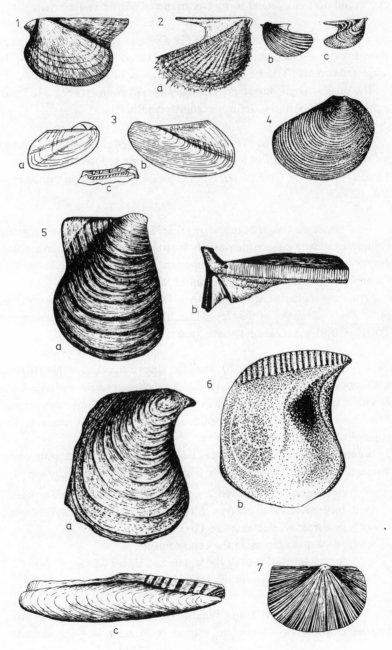

Pterinea (Fig. 81, 1): Inequivalve, oblique, right valve flattened, left valve convex; with small anterior auricle; ligament accommodated in several grooves parallel with the hinge line; two or three oblique hinge teeth below the umbo, and posterior lateral teeth. Occurrence: Ordovician-Devonian.

Pteria (syn. *Avicula*) (Fig. 81, 2): Oblique oval shell, slightly inequivalve; short anterior auricle, long posterior auricle. One tooth on either side of the hinge. Occurrence: Triassic-Recent. Recent forms inhabit warm seas.

The small strongly curved *Pteria (Rhaetavicula) contorta* (Fig. 81, 2b) is a guide fossil in both the Germanic and Alpine Rhaetian.

Gervillia (Fig. 81, 3): Elongated and twisted shell. A few lateral teeth below the umbo. Ligament lodged in few transverse grooves. Occurrence: Triassic-Cretaceous. *Gervillia (Hoernesia) socialis* is frequent in the Germanic Muschelkalk.

Posidonia (syn. *Posidonomya*) (Fig. 81, 4): Thin shell, inequilateral, oblique, oval, with concentric ribs. Short straight hinge line. Auricles not well developed. *Posidonia becheri* is an important guide fossil for the Lower Carboniferous (Kulm facies). Occurrence: Silurian-Jurassic.

Steinmannia bronnii, which used to be known as *Posidonia* and which occurs in the *Posidonia* shales of the Upper Lias, is very similar, but is restricted to the Toarcian. Both genera seem to have lived as pseudoplankton.

Inoceramus (Fig. 81, 5): Shell egg-shaped to rounded, inequilateral; with concentric ribbing; numerous small ligament pits along the hinge line. Prismatic layer very thick, often disintegrated, in Upper Cretaceous representatives. *Inoceramus* is an important guide fossil for the Upper Cretaceous. Occurrence: Jurassic-Cretaceous.

Numerous genera and species have been distinguished which cannot be discussed in detail. Stratigraphically important *Inoceramus* species are:

1. *Inoceramus cuvierii* (syn. *I. schloenbachi*) in the Upper Turonian;
2. *Inoceramus lamarcki* (syn. *I. brongniarti*) in the Middle Turonian;
3. *Inoceramus labiatus* in the Lower Turonian;
4. *Inoceramus crippsi* in the Cenomanian;
5. *Inoceramus sulcatus* in the Middle and Upper Albian;
6. *Inoceramus dubius* in the Upper Lias.

Isognomon (syn. *Perna, Pedalion*) (Fig. 81, 6): Inequilateral shell, terminal umbo; straight hinge line, without teeth, numerous ligament pits;

byssal notch in the right valve; large subcentral posterior adductor scar.
Occurrence: Triassic–Recent.

Daonella (Fig. 81, 7): Thin shell, similar to *Posidonia* but without
auricles; adductor scar bounded by two ridges radiating from the umbo. Occurrence: Triassic. *Halobia*, also Triassic, is similar, with an anterior auricle.

In the following seven genera of so-called Isodonta, the hinge consists of
symmetrically arranged teeth and sockets either side of the ligament pit.

Pecten (scallop) (Fig. 82, 1): Equilateral shell, but inequivalve, right
valve convex, left valve flattened, monomyarian, integripalliate; straight hinge
line with two roughly equal auricles at front and rear; ligament in a central pit.
Shallow byssal notch. Pronounced radiating ribs. Swimming movements by
forcibly bringing the two valves together. Very diverse genus. Occurrence:
Cretaceous–Recent.

Chlamys (Fig. 82, 2): Similar to *Pecten* but anterior auricles larger than
the posterior ones, with deep sinus for the byssus. Occurrence: Triassic–Recent.

Neithea (Fig. 82, 3): Inequivalve; right valve strongly convex, left valve
flattened or concave. Sculpturing consists of alternating strong and weaker
radiating ribs. Occurrence: Cretaceous.

Buchia (syn. *Aucella*) (Fig. 82, 4): Inequivalve; left valve convex, right
valve planar. Oblique, thin shell, with robust concentric ribs; right valve with
small anterior auricle; short hinge line. Occurrence: Jurassic–Lower Cretaceous,
especially widespread in the Boreal zone.

Spondylus (Fig. 82, 5): Thick shell, inequivalve, equilateral, integripalliate; attached by the larger (right) valve which has a prominent area; ligament
in central pit; radial ribs furnished with spines. Occurrence: Jurassic–Recent.

Plicatula (Fig. 82, 6): Similar to *Spondylus* but usually smaller; divergent hinge teeth, tiny area; surface with growth lamellae which may protrude as
spines. Occurrence: Triassic–Recent.

Lima (Fig. 82, 7): Shell more or less equivalve; elongated towards the
front (protruncate); hinge line more or less toothless, with small anterior and
posterior auricles. Adductor scar well to the rear; gaping slightly at the front.

Fig. 82. Pteriomorphia (Order Pterioida): (1) *Pecten* (× 0.1); (2) *Chlamys* (× 0.2); (3) *Neithea* (× 0.25); (4) *Buchia* (× 0.5); (5) *Spondylus* (a, b: × 0.35; c: × 0.5); (6) *Plicatula* (× 0.6); (7) *Lima* (a, b: × 0.6; c, d: in life position, × 0.3).

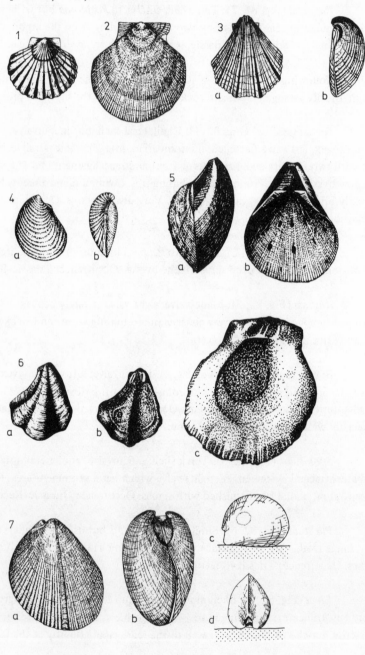

Capable of swimming movements like *Pecten*. Present-day forms rest on their front edge. Sculpturing smooth (subgenus *Plagiostoma*) to strongly ribbed (subgenus *Ctenostreon*). Occurrence: of the genus, Triassic-Recent; of the Family Limidae, Carboniferous-Recent.

The following four genera, which are also monomyarian, are 'oysters' in the wider sense, attached by their left, usually more convex, valve. The ligament is lodged in a triangular pit; no hinge teeth; only one adductor muscle scar placed centrally.

Ostrea (oyster) (Fig. 83, 1): Right valve flattened; shell with lamellar structure. Occurrence: Cretaceous-Recent.

Lopha (syn. *Alectryonia*) (Fig. 83, 2): Shell similar to that of *Ostrea* but folded radially in a zig-zag fashion. This form should probably only be assigned subgeneric status. Occurrence: Upper Triassic-Recent.

Gryphaea (Fig. 83, 3): Shell strongly inequivalve, with strongly convex and incurved left valve which is fixed in the juvenile stage only. The right valve

Fig. 83. Pteriomorphia (Order Pterioida): (1) *Ostrea* (× 0.4); (2) *Lopha* (× 0.5); (3) *Gryphaea* (× 0.5); (4) *Exogyra* (× 0.3).

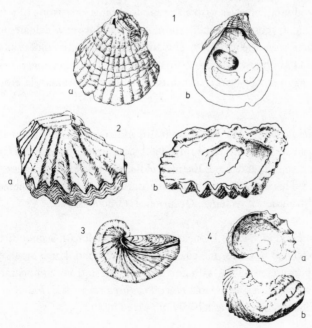

Fig. 84. Palaeoheterodonta: (1) *Schizodus* (× 0.5); (2) *Trigonia* (× 0.35); (3) *Myophoria* (× 1); (4) *Unio* (× 0.4); (5) *Anthracosia* (× 0.5).

forms a flat lid. This genus is probably heterogeneous (iterations). Occurrence: Lower Jurassic–Recent.

Exogyra (Fig. 83, 4): Small to fairly large shells with peripheral spiral umbo. Occurrence: Middle Jurassic–Recent. This genus provides a number of Cretaceous guide fossils.

4.   Subclass: Palaeoheterodonta

'Schizodonta', bivalves with an actinodont hinge consisting of a few cardinal teeth and at most some undifferentiated lateral teeth which are not distinctly separate from the cardinals. This subclass includes the oldest known bivalves (see p. 112), the Actinodonta, Unionidae and Trigoniidae, representatives of which may perhaps be the ancestors of the Pteriomorphia, Heterodonta and others.

Lyrodesma (Fig. 76a): Small equivalve and concentrically striated shell. The hinge consists of 5–9 crenulated ridge teeth converging towards the umbo. The pallial line is slightly indented; the rear adductor muscle scar is larger than the anterior one. This genus is regarded as the starting point for the 'Schizodonta' because of the structure of its hinge. Occurrence: Ordovician.

Schizodus (Fig. 84, 1): Elongated triangular shell with a smooth surface with slightly extended posterior margin. Opisthogyral umbo. Large indented triangular tooth, not crenulated, with one lateral tooth in front and behind. Occurrence: Carboniferous–Permian. *Schizodus obscurus* is often found in Upper Permian (Zechstein) sediments as a badly preserved mould.

Trigonia (Fig. 84, 2): Triangular, thick-walled, strongly sculpted shell with opisthogyral anterior umbones. The posterior area, which is divided off by a ridge, is sculpted differently from the rest of the shell. The hinge teeth are crenulated, the triangular tooth of the left valve deeply indented. The adductor scars are backed up by ridges. Occurrence: of the present (narrowly defined) genus, Triassic–Cretaceous; of the Family Trigoniidae, Triassic–Recent.

In Central Europe *T. navis* and *T. costata* are guide fossils for the Lower and Upper Dogger respectively.

Myophoria (Fig. 84, 3): Usually triangular; the posterior portion is divided off by one or several ridges and usually has different sculpturing. Simple or bifid hinge tooth, sometimes crenulated. Occurrence: of the genus, Triassic; of the Family Myophoriidae, Devonian–Triassic. Numerous species in the Germanic and Alpine Triassic.

Unio (Fig. 84, 4): Very variable in shape; large thick-shelled inhabitants of stream and rivers. External ligament. Hinge generally with large grooved and weak anterior tooth on the right, and with two grooved diverging teeth and two lateral teeth on the left. The structure of the hinge is very variable. Occurrence: Jurassic–Recent; nowadays divided into numerous genera.

The fresh-water mussel *Anodonta* with its virtually toothless hinge but thinner shell is very similar (in existence since the Upper Cretaceous).

Anthracosia (Fig. 84, 5): Very variable shape, usually small; dimyarian, integripalliate; the dentition often consists of one hinge tooth on each side. This genus is a characteristic inhabitant of the Carboniferous 'coal' swamps where it sometimes occurs in lumachelles, a rock consisting mainly of mollusc remains. Occurrence: Upper Carboniferous (Westphalian).

5.   Subclass: Heterodonta
This subclass includes bivalves with heterodont hinges which are predominantly free-living. They have eulamellibranch gills. Phylogenetically speaking, they are the youngest and, at present, the most dominant group, which is still undergoing rapid development.

1.   Order: Veneroida
This is the most diverse bivalve order; it includes mostly equivalve and isomyarian forms which are adapted to a whole range of different biotopes and in some cases still in the process of developing their full potential.
Examples:

Lucina (Fig. 85, 1): Medium-sized to large, rounded, integripalliate bivalves with two cardinal teeth and one or two lateral teeth, which may be reduced, for each valve. Sculpturing smooth or with concentric ribbing. Occurrence: of the genus, Upper Cretaceous-Recent; of the Family Lucinidae Silurian-Recent.

Chama (Fig. 85, 2): Usually fairly small, rounded shell with the left valve fixed; robust concentric or radiating sculpturing; prosogyral umbones; dimyarian; left, two hinge teeth; right, one hinge tooth. The 'inverse' genus *Pseudochama* is fixed by its right valve and accordingly has two hinge teeth on the right and only one on the left. Occurrence: Lower Tertiary-Recent.

Fig. 85. Heterodonta (Order Veneroida): (1) *Lucina* (× 0.4); (2) *Chama* (× 1); (3) *Cardita* (× 0.5); (4) *Astarte* (× 0.8); (5) *Cardinia* (× 0.5); (6) *Cardium* (× 0.7).

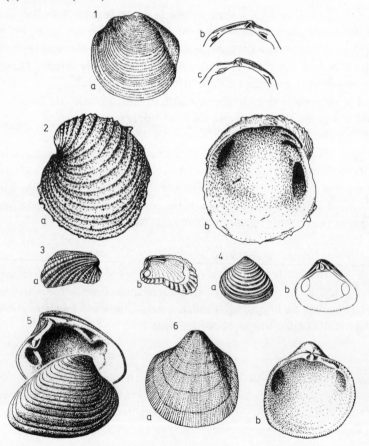

Cardita (Fig. 85, 3): Large thick-walled shell with radial ribs; strongly convex; prosogyral umbones moved well to the anterior; external ligament; two hinge teeth left, three right; reduced laterals. Anterior adductor scar on a platform. Inner margin crenulated. Occurrence: of the genus, Palaeocene-Recent; of the Family Carditidae, Devonian-Recent.

The same family includes *Venericardia*, which differs from *Cardita* by the possession of a small lateral tooth at the front and rear of the right valve. Occurrence: Palaeocene-Recent.

Astarte (Fig. 85, 4): Small thick-walled shell with concentric ribbing and external ligament; two cardinal teeth on either side; pedal impression above the anterior adductor scar. Present-day forms predominantly live in cold seas. Occurrence: of the genus, Jurassic-Recent; the Family Astartidae, Devonian-Recent.

Cardinia (Fig. 85, 5): Medium-sized, elongated, thick-shelled with lunule; anterior umbo; a single cardinal tooth in the right valve, which may also be absent; anterior lateral teeth short, posterior ones long, ending in tubercles. Occurrence: Triassic–Jurassic. (*'Thalassites* beds' in the Lias of Southern Germany, derived from the previous name *Thalassites*.)

Cardium (cockle) (Fig. 85, 6): Equivalve and often equilateral shell, rounded, usually with radiating ribs; external ligament, opisthodetic; hinge with two opposite conical cardinal teeth and a lateral tooth to the front and the rear, two at the front in the right valve. Very mobile, capable of jumping with the aid of the long finger-like foot. Occurrence: the genus, Miocene-Recent; of the Superfamily Cardiacea, Triassic-Recent.

Some of the numerous genera have migrated into brackish-water habitats. During this process they have undergone several modifications such as atrophy of the hinge, the development of a pallial sinus, gaping valves, etc. (e.g. *Lymnocardium*).

Tridacna (Fig. 86, 1): Sometimes gigantic shells with radial ribbing and a byssal notch in front of the umbo, serrated margin; external ligament; opisthodetic; one cardinal tooth in each valve; one posterior lateral tooth on the left, two on the right. *Tridacna* inhabits warm oceans and lives between corals, attached by its byssus threads. It is the largest present-day bivalve, and lives in symbiosis with Zooxanthellae. Occurrence: Lower Tertiary-Recent.

Mactra (Fig. 86, 2): Small thin-shelled triangular burrowing bivalves with pallial sinus and internal ligament lodged in a large ligament pit (isolated

Fig. 86. Heterodonta (Order Veneroida): (1) *Tridacna* (× 0.1); (2) *Mactra* (× 0.3); (3) *Solen* (× 0.2); (4) *Macoma* (× 0.4); (5) *Donax* (S, pallial sinus; × 0.4); (6) *Scrobicularia* (× 0.5); (7) *Dreissena* (× 0.5); (8) *Congeria* (× 0.15).

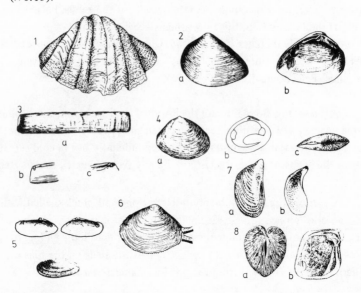

from the external ligament) below the umbo; in front, inverted V-shaped cardinal tooth in the left valve with a corresponding socket in the right valve. On the left side one, on the right side two robust lateral teeth. Occurrence: of the genus Eocene–Recent; of the family, Upper Cretaceous–Recent.

In the subgenus *Spisula* the internal ligament is not separated from the external ligament.

Solen (syn. *Vagina*; grooved razor shell) (Fig. 86, 3): Cylindrical shell, open at both ends, umbo close to the anterior end; weak hinge with a cardinal tooth on each side. Occurrence: Eocene–Recent.

Ensis is similar but more flattened, umbo close to the posterior margin, two cardinal teeth left and right. Occurrence: Eocene–Recent.

Razor shells are able to burrow very rapidly with the aid of their powerful burrowing foot. They live in deep vertical passages which they are able to descend and ascend. Occurrence: of the Superfamily Solenacea, Lower Cretaceous–Recent.

Tellina (tellin): Thin-walled oval shell with a blunt fold extending from the umbo to the posterior border; on the right two cardinal teeth of

which the posterior one is bifid; one anterior lateral tooth. Deep pallial sinus.
Tellins tend to inhabit warm seas. Occurrence: of the genus, Tertiary–Recent;
of the diverse superfamily, which also includes the following four genera,
Triassic–Recent.

In the related genus *Macoma* (Fig. 86, 4), which was previously regarded as
a subgenus, the lateral teeth are reduced. *M. balthica* (Baltic tellin) is a Boreal
coastal form which is widespread in the North Sea and western Baltic Sea.
Occurrence: Eocene-Recent.

Donax (Fig. 86, 5): Small to medium-sized shell but more robust than
Tellina; opisthogyral, anterior end longer than posterior end, external ligament,
crenulated margin, pallial sinus.

D. vittatus is widespread in the North Sea but already sparse in the Kattegat.
Occurrence: Eocene–Recent.

Scrobicularia (Fig. 86, 6): Oval medium-sized shell; two cardinal teeth,
no lateral teeth, ligament external and as a resilium in an internal chondrophore.
Syndosmia is similar but possesses lateral teeth.

These bivalves search the surface for food with the aid of long siphons; they
live buried in dense aggregates near the coast, including that of the western
Baltic Sea.

Dreissena (zebra mussel) (Fig. 86, 7): Shell similar to that of *Mytilus*
but without mother-of-pearl layer; in the region of the umbo a short platform
acts as a myophore for the anterior adductor muscle. Attached by byssus threads.

The Recent *D. polymorpha*, a fresh-water inhabitant, has spread out all over
Europe, having started from the south-east at the beginning of the nineteenth
century. It is now one of the most abundant fresh-water bivalves. Occurrence:
of the genus and the Family Dreissenidae, Eocene–Recent.

Congeria (Fig. 86, 8): Similar to the closely related *Dreissena*, but shell
square with incurved umbones. Below the anterior adductor muscle platform
there is a small ledge for the support of the pedal muscle. Occurrence: Oligocene-
Pliocene, especially abundant in the south-east European *Congeria* beds.

Arctica (syn. *Cyprina*) (Fig. 87, 1): Large thick-walled shell, egg-shaped,
strongly convex with concentric striae; integripalliate. Hinge of the 'cyrenoid'
type: right valve with three cardinal teeth and two anterior and posterior teeth,
left also with three cardinal teeth, but one anterior and two posterior teeth. The
teeth have a tendency to radiate from the beak. Occurrence: Lower Cretaceous-

Fig. 87. Heterodonta (Order Veneroida): (1) *Arctica* (× 0.4); (2) *Corbicula* (× 0.7); (3) *Venus* (× 0.4); (4) *Tapes* (× 0.4); (5) *Petricola* (× 0.6); (6) *Mya* (*Arenomya,* × 0.4).

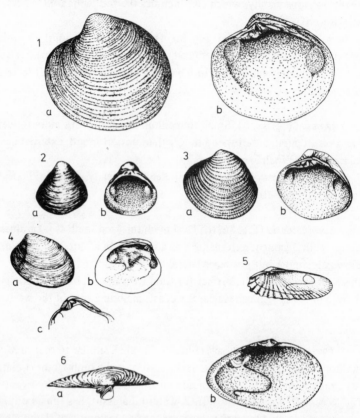

Recent; nowadays inhabits cold seas. *A. islandica* is the guide fossil for the beginning of the Pleistocene in the Mediterranean.

Corbicula (syn. *Cyrena*) (Figs. 76g and 87, 2): Usually medium-sized bivalves with well developed periostracum; integripalliate; external ligament; complete hinge with single lateral teeth in the left valve and double ones in the right valve. Modern forms inhabit brackish water or fresh water, mainly in estuaries of the warm zones; fossil forms also inhabited marine biotopes. Occurrence: of the genus, Cretaceous–Recent; of the family, Jurassic–Recent.

Pisidium: Similar to *Corbicula* but very small, with internal ligament but with no more than two hinge teeth on each side. Recent forms are inhabitants of fresh water. Occurrence: Cretaceous–Recent.

Venus (Fig. 87, 3): Oval equivalve shell usually with concentric sculpturing; pronounced ligament nymphs (fulcra) for the external ligament; weak pallial sinus; hinge with three cardinal teeth on each side but no laterals. Occurrence: of the genus, Oligocene-Recent; of the Family Veneridae, Lower Cretaceous-Recent.

Tapes (Fig. 87, 4): Also a member of the Family Veneridae. Shell oblique, oval, commonly with concentric ribbing; hinge with three diverging, often bifid, cardinal teeth; no lateral teeth; deep pallial sinus; long straight lunule. Occurrence: Miocene-Recent.
Paphia senescens of the subgenus *Paphia* is a guide fossil for the Eem interglacial period.

Petricola (Fig. 87, 5): Shell elongated and oval with radiating ribs; deep pallial sinus; two cardinal teeth on the right, three on the left; no lateral teeth; anterior ribs coarse. *Petricola* tend to bore into corals, rocks and mud by mechanical means. *P. (Petricolaria) pholadiformis* was transported to England from North America in about 1880 and has since spread as far as the western Baltic Sea by larval transport. It bores into peat, silt and sand. Occurrence of the genus *Petricola*: Eocene-Recent.

Mya (Figs. 76j and 87, 6): Large, oval, almost toothless (desmodont) bivalves with deep pallial sinus; internal ligament on a chondrophore in the left valve, with a corresponding pit in the right valve; lives deeply buried. Nowadays this genus is found mainly in muddy areas and in the estuaries of cold seas. *Mya (Arenomya) arenaria* (sand gaper) in the Northern Atlantic; *Mya (Mya) truncata* in the northern and Polar oceans. Occurrence of the genus: Oligocene-Recent.

Pholas (common piddock) (Fig. 88, 1): Shell elongated, oval, gaping at front and rear; hinge line toothless, commonly folded back at the umbo; deep pallial sinus; without a ligament. A so-called styloid apophysis projects freely into the shell interior from each umbo; it serves for inserting the pedal muscle. Anterior and posterior sculpturing different. *Pholas* bores mechanically into wood, peat and also rock, with the foot serving as a hold. The two adductor muscles work differently so that the valves are rotated. Occurrence: Cretaceous-Recent.

Teredo ('shipworm') (Fig. 88, 2): Very small shell, gaping on both sides, surrounds only the anterior part of the body. *Teredo* bores mechanically into

Fig. 88. Heterodonta (Order Veneroida): (1) *Pholas* (× 0.4); (2) *Teredo*, left valve: (a) exterior, (b) interior (× 3.5); (3) *Megalodon* (× 0.4); (4) *Conchodon* (× 0.2). (Order Hippuritoida): (5) *Diceras* (× 0.5).

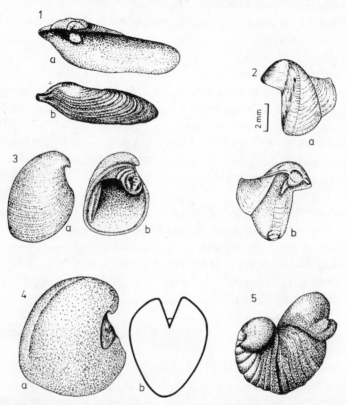

wood, digests the resulting wood filings and lines the cavity with a secondary calcareous layer. The ends of the long passages it makes (up to 1.2 m) may be closed off with calcareous plugs. Occurrence: Cretaceous–Recent.

Megalodon (Fig. 88, 3): Shell strongly convex, rounded, triangular, thick, equivalve; wide hinge plate with two cardinal teeth on each side; posterior adductor scar on a projecting ridge, anterior one small and well sunken. Occurrence: Devonian–Triassic.

The subgenus *Neomegalodon* has similar, slightly larger representatives. The posterior right tooth is commonly divided by a longitudinal groove. Restricted to the Upper Triassic.

Conchodon (Fig. 88, 4): Very large, thick shells with pronounced prosogyrally coiled umbo. The valves have a sharp keel at the rear. Cross-sections of *C. infraliasicus* with diameters of up to 18 cm are sometimes found in enormous numbers in the Rhaetian of the Alps (Dachstein limestone); they are known as Dachstein bivalves.

2.   *Order: Hippuritoida*

     The hippuritids in the narrower sense (Superfamily Hippuritacea) are particularly remarkable on account of their shell, which is reminiscent of corals and Richthofeniidae. From the Late Jurassic to the Latest Cretaceous they were among the characteristic inhabitants of neritic warm-water zones between southern Sweden and Madagascar.

     The hippuritids are inequivalve and fixed by the right or left valve; they are rarely free-living. The hinge of the unattached valve has two teeth and a socket, that of the attached valve one tooth and two sockets, with the exception of *Diceras* which is fixed by its right valve and always has two teeth in the right valve. The two adductor scars are either on the shell wall or on special myophores.

     Examples:

     Diceras (Figs. 76e and 88, 5): Both valves with strongly coiled umbones; inside of shell reminiscent of megalodontids. Occurrence: Upper Jurassic.

     Requienia (Fig. 89, 1): Inequivalve, left valve fixed, right valve a flat coiled lid; hinge as in *Diceras* with two teeth in the right (lid) valve. Occurrence: Lower to Upper Cretaceous.

     Caprina (Fig. 89, 3): Inequivalve, right valve fixed and elongated, left valve coiled; mantle canals in both valves. Occurrence: Lower to Upper Cretaceous.

     Hippurites (Fig. 89, 2): Right valve fixed, large, with longitudinal ribs, tall, conical with three vertical grooves; left valve lid-like, porous, with central umbo, two openings (osculi); inner layer (hypostracum) of the right valve forms three folds. Occurrence: Upper Cretaceous (Turonian to Maastrichtian).

     Radiolites (Fig. 89, 4): Inequivalve, right valve fixed, conical, left valve lid-like and flat; right valve without folds; outer shell layer (ostracum) blistery-cellular. Occurrence: Upper Cretaceous (Cenomanian to Maastrichtian).

6.   Subclass: Anomalodesmata

     This small group of eulamellibranch or septibranch bivalves is more or less isomyarian with a thickened or incurved hinge line and no teeth. It includes

Fig. 89. Heterodonta (Order Hippuritoida): (1) *Requienia*: (a) lateral view (×0.25), (b) left (attached) and right (free) valves, hinge area (×0.35). The numbers indicate the hinge teeth. (2) *Hippurites*: (a) view of the front with the three characteristic longitudinal grooves: (b) view of the right (fixed) valve (C, column (buttress); L + D, ligament and dental supports; MP, muscle pit; MS, muscle scar; S, socket); (c) transverse section of the right valve (1, 3, sockets; 2, tooth; PM, pit for the rear myophore; C, column (buttress)); (d) left valve ('lid') with the front and rear tooth and (on the far left) the rear myophore (all ×0.3). (3) *Caprina*: (a) lateral view of both valves (×0.05); (b) section of the right (attached) valve (×0.25). (4) *Radiolites* (×0.5).

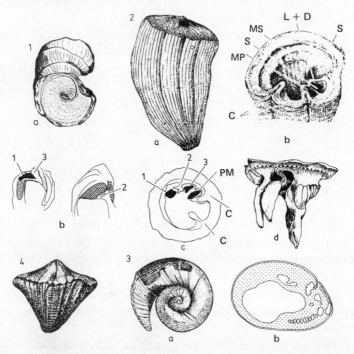

some of the so-called Desmodonta. Its representatives are mostly burrowing forms.

The body plan of these thin-shelled burrowing bivalves is represented by numerous genera which usually differ in their sculpturing; their internal structure is hardly known.

Examples:

Grammysia (Fig. 90, 1): Shell elongated obliquely, with concentric striae or wrinkles, deep lunule below the umbo; integripalliate; hinge without teeth; two oblique grooves from the umbo to the ventral border. Occurrence: Devonian.

Fig. 90. Anomalodesmata: (1) *Grammysia* (× 0.6); (2) *Goniomya* (× 0.6); (3) *Pleuromya* (× 0.6); (4) *Poromya* (× 2).

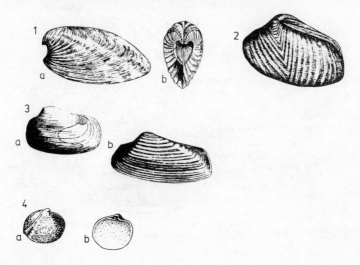

Goniomya (Fig. 90, 2): Similar to *Grammysia* but has V-shaped ribs. Occurrence: Jurassic–Lower Tertiary.

Pleuromya (Fig. 90, 3): Thin medium-sized shell, no teeth, pronounced nymphs for the opisthodetic external ligament; sinupalliate; fine blisters on the surface; valves gaping at the rear. Occurrence: Triassic–Lower Cretaceous.

Poromya (Fig. 90, 4): With septibranch gills. The Recent representatives of this small group are inhabitants of the deep sea, and have a special pumping mechanism for food intake. Occurrence: Cretaceous–Recent.

6.   Class: Cephalopoda
        The cephalopods are the most highly developed class of molluscs and the only class to have produced active swimmers with large 'brains' and efficient sense organs, particularly eyes, although a few bivalves like *Pecten* and *Cardium* also have eyes and occasionally swim. Cephalopods have always lived in exclusively marine habitats. They include the largest living invertebrates such as *Architeuthis* (up to 22 m long including the arms).
        Their characteristic organs, namely the funnel and arms, are modifications of the foot. Recent cephalopods usually have strong jaws, but the radula is less diverse than in the gastropods (Fig. 91).
        Fossil cephalopods have provided the most important macro-guide-fossils from the Devonian to the end of the Cretaceous.

Fig. 91. Median section of a Recent nautilus.

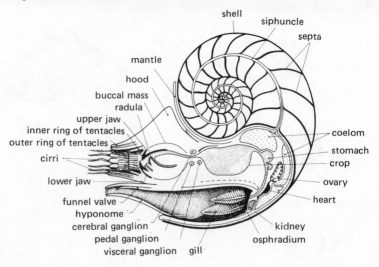

Summary of Recent cephalopods

Among the Recent cephalopods two groups are distinguished which differ considerably in terms of diversity:

A. Tetrabranchiata (Ectocochlia): cephalopods with two pairs of gills, numerous arms and an external shell. *Nautilus* ('pearly nautilus') is the only representative of this group.

B. Dibranchiata (= Coleoidea, Endocochlia): cephalopods with one pair of gills. The shell is inside the soft body, if and when it is developed at all. This group includes:

1. The 'Decapoda' (squids, Decabrachia with the Orders Teuthida and Sepiida) with ten arms of which two are differentiated into tentacles.

2. The 'Octopoda' (octopuses, Octobrachia) with eight arms and a sac-like body.

Summary of fossil cephalopods

Among the fossil cephalopods two large groups were formerly distinguished:

A. Forms with an external shell (Ectocochlia): (1) nautilids, with simple septa; (2) ammonoids, with complicated folded septa.

B. Forms with internal shells (Endocochlia): belemnites and related forms.

On account of their common ancestry and similar radulae, the ammonoids and the coleoids are established as closely related sister groups, as opposed to the primitive group of the archaic *Nautilus* relatives. In this book the former are thus combined as the Infraclass Neocephalopoda and the latter as the Infraclass Palcephalopoda (= 'nautilids').

1.   Infraclass: Palcephalopoda (nautilids in the wider sense)
        The first three subclasses of the cephalopods are here combined as the Palcephalopoda: the Nautiloidea, Endoceratoidea and Actinoceratoidea.

The morphology of fossil Palcephalopoda (nautilids)

The fossil relatives of the present-day *Nautilus* also had an external shell. The animals lived in the most recently formed part of the shell, the so-called body chamber, which was open to the outside. The older part of the shell, the phragmocone or chambered part, is divided into numerous chambers by dividing walls known as septa. The very first chamber to be formed is known as the protoconch (initial or apical chamber). A membranous tube, the siphuncle, traverses all the chambers from the protoconch to the body chamber.

The significance of the siphuncle

Studies of Recent cephalopods with shells have revealed that the chambers contain a gas mixture resembling atmospheric air without oxygen, i.e. consisting mainly of nitrogen, with a gas pressure of about 0.8 atmospheres. Newly formed chambers are initially filled with fluid. Only when the new septum is sufficiently strong to resist the water pressure will the siphuncle pump the water out, sometimes against a substantial external pressure. The siphuncle contains arterial and venous blood vessels. Its tissues are capable of selectively excreting water and salts, rather like kidney tissue.

The shells of fossil nautilids may be rolled into a logarithmic spiral like the *Nautilus*, but most of them, especially the geologically oldest ones from the Cambrian to Silurian, are more or less straight. The siphuncle may occupy different positions varying from central to peripheral.

Using the present-day *Nautilus* as a reference, the siphuncle is regarded as a ventral organ. Accordingly, the following coiled forms are distinguished: (*a*) endogastric (siphuncle on the inside of the whorl); (*b*) exogastric (siphuncle on the outside of the whorl).

Where the siphuncle penetrates the dividing wall (septum) a calcareous septal neck is formed round the siphuncle. In the area between the septal neck and

Fig. 92. Longitudinal section of part of an actinoceratoid shell. The stippled areas represent the areas filled by primary calcareous secretions during the lifetime of the animal.

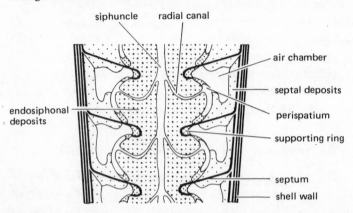

the next septum, the siphuncle is surrounded by a membranous siphonal layer (= connecting ring, ectosiphon) (Fig. 92).

The siphuncle may be partly or completely filled with endosiphonal calcareous secretions (Fig. 93).

According to the structure of the siphuncle, stenosiphonate (with a simple narrow siphuncle) and eurysiphonate (with a wide siphuncle) forms are distinguished.

In addition to the siphuncle, the air chambers may also be more or less extensively filled with calcareous deposits.

The following terms are commonly used to describe the shape of the shell:

orthocone: straight

Fig. 93. Diagrammatic representation of the endosiphonal calcareous secretions (stippled) in the nautilids. The top row shows transverse sections, the bottom row longitudinal sections.

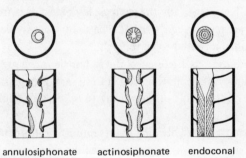

annulosiphonate   actinosiphonate   endoconal

cyrtocone: slightly curved
gyrocone: coiled into a loose spiral
nautilicone: completely coiled

The orthocone and cyrtocone shells are either brevicone (short, squat) or longicone (elongated, slender).

In contrast to the jaws of all other Recent cephalopods, the jaws of the Recent *Nautilus* are partly calcified. The radula has 13 teeth per transverse row and is thus fairly wide ('latiradulate').

With very few exceptions, the dividing walls (septa) are simple and at most slightly arched. This applies particularly to the marginal parts which are fused with the outer shell wall (the so-called suture line).

An indentation in the initial chamber of some nautilids which sometimes looks like a healed injury is known as cicatrix ('scar'). It does not occur in the ammonoids. Its origin and function are unknown, but they are probably linked with the initial part of the siphuncle, the caecum.

In the Upper Ordovician the Palcephalopoda achieved their maximum size with up to 9 m in the endoceratids. With the exception of the Recent *Architeuthis*, the cephalopods have never surpassed this size.

In terms of the number of species, their climax of development was in the period between the Upper Ordovician and the Lower Devonian.

Coiling first began in the Ordovician and prevailed until the Triassic, with the concomitant decline of the orthocone, cyrtocone and gyrocone forms. Only coiled nautilids are known in the post-Triassic Period.

Palaeoecology and biostratonomy

As the only Recent form, the present-day *Nautilus* is predominantly distributed in the Indo-Pacific area. The shells may have carried over great distances after death. In experiments, even slightly damaged empty *Nautilus* shells floated for about one month. In fossil forms the length of the body chamber is important. An empty *Endoceras* will float provided the body chamber is not longer than one-quarter of the phragmocone (less than one-half in *Orthoceras*). Shells with evolute coiling do not float as easily as strongly involute shells; strongly compressed forms sink. Shells covered with epizoic serpulids and bivalves cannot normally have drifted after death.

Nautilus is caught in traps down to depths of 400–500 m. At night, schools of *Nautilus* migrate to the coast. They feed mainly on small crabs and carrion. Locomotion is by swimming with jet propulsion, not by crawling. The protective colouring on the dorsal side is very effective; remains of colouring have also been found in fossil specimens. If the fossil nautilids also moved by a recoil mechanism, which is not certain in all cases, then the funnel and body chamber must have been in a roughly horizontal position: otherwise one could only imagine a slow creeping movement or an equally slow vertical movement. Early ellesmeroceratids with very narrowly spaced septa will certainly have been part of the mobile epibenthos. In their case, equilibrium in the water would have been guaranteed by a little buoyancy in the chambers.

The following shell types predominate:

1. More or less complete coiling, stability provided by ventral position of the body chamber.
2. Heavy cameral and siphonal deposits, particularly in the ventral and apical regions: stable horizontal position (Fig. 94).
3. Ontogenetic loss of older parts of the shell by truncation (decollation), air chambers in the dorsal part of the teleoconch: stable horizontal position.
4. Formation of brevicone shell: stable vertical position because the heavy body is in the body chamber at the bottom.

Any preserved colouring is distributed accordingly: on one side only in longicone shells and more or less evenly all round in brevicone shells.

Fossil nautilids may have had ink sacs, but there is still no unequivocal evidence to this effect.

Fig. 94. Longitudinal sections of various nautilid shells of the Lower Palaeozoic. Stippled areas, body chamber and siphuncle; black areas, calcareous secretions inside the chambers and siphuncle.

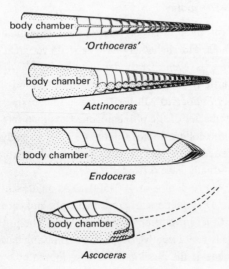

body chamber

'Orthoceras'

body chamber

Actinoceras

body chamber

Endoceras

body chamber

Ascoceras

Calculations of the maximum diving depths, based on the pressure resistance of the siphuncle and the shell, have yielded the following values:

Endoceras	450 m
Plectronoceras	250 m
Orthoceras	400 m
Nautilus	800 m
Bactrites	550 m
Spirula	1600 m

These are maximum values, and the actual depths inhabited will probably only have been about half the values shown, as illustrated by the example of *Nautilus*. Most fossil nautilids were thus inhabitants of the shallow shelves. On the other hand, the water could not be too shallow, otherwise the large forms especially would have been jeopardised by far-reaching wave movements. A planktonic lifestyle may have been preferred by small forms such as the bactritids, similar to present-day *Spirula*. Larger forms were probably part of the benthos.

1.   Subclass: Nautiloidea

Since the (? Lower) Upper Cambrian. A very diverse central group; commonly straight at first; geologically younger forms are strongly involute. The position of the siphuncle is variable; it is thin, orthochoanitic or cyrtochoanitic (Fig. 95).

The status of the Lower Cambrian Genus *Volborthella* which tends to be assigned to the hyolithids or to the annelids by some authors is very doubtful (Fig. 96*A*, 1). In the Upper Cambrian small forms (Order Ellesmerocerida) such as *Plectronoceras* (Fig. 96*A*, 2) and *Knightoconus* predominate. Explosive development ensued in the Lower Ordovician.

The Orthocerida have straight shells. Occurrence: Ordovician-Triassic. *Orthoceras* (Fig. 96*A*, 3), *Michelinoceras*. Their division from similar genera and the time of their occurrence have still not been satisfactorily resolved.

Fig. 95. The most important types of cephalopod siphuncle.

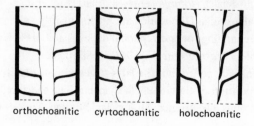

orthochoanitic   cyrtochoanitic   holochoanitic

Fig. 96*A*. Nautiloidea: (1) *Volborthella* (× 5); (2) *Plectronoceras* (× 1); (3) *Orthoceras* (× 0.5); (4) *Ascoceras* (× 0.4); (5) *Oncoceras* (a: × 0.5; b: × 2); (6) *Gomphoceras* (× 0.3).

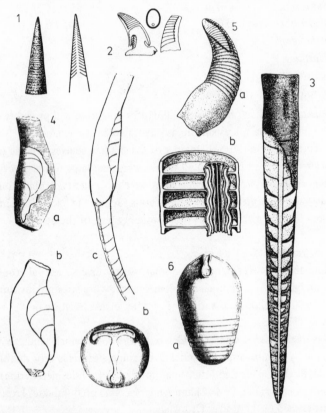

The Ascocerida discarded the oldest, orthochoanitic part of the shell (so-called decollation); the remaining part had a cyrtochoanitic siphuncle and septa extending well to the front. Occurrence: Ordovician–Silurian. *Ascoceras* (Figs. 94 and 96*A*, 4; Silurian).

The Oncocerida are exogastric cyrtocones, often with a narrow aperture and actinosiphonate deposits. Occurrence: Middle Ordovician–Lower Carboniferous. Examples: *Oncoceras* (Ordovician), *Gomphoceras* (Silurian) (Fig. 96*A*, 5, 6).

In contrast, the externally similar Discosorida are predominantly endogastric with thick complex connecting rings or endosiphonal deposits and often with a narrow aperture. Example: *Phragmoceras* (Fig. 96*B*, 3; Silurian).

The Tarphycerida are initially coiled, but the later stages are uncoiled to varying degrees, with the position of the siphuncle changing. The aperture has a conspicuous hyponomic sinus. Occurrence: Ordovician–Silurian. This group

Fig. 96*B*. Nautiloidea: (1) *Lituites* (×0.27); (2) *Ophioceras* (×1);
(3) *Phragmoceras* (×0.2); (4) *Nephriticeras* (×0.44); (5) *Aturia* (×0.13);
(6) *Germanonautilus* (×0.2); (7) *Trochoceras* (×0.2).

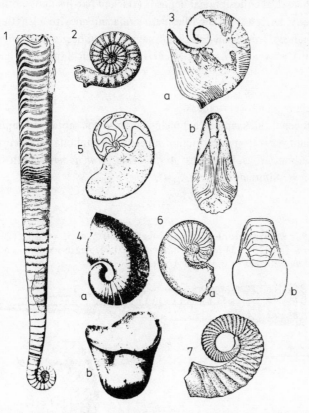

includes the genera *Lituites* and *Ophioceras* (Fig. 96*B*, 1, 2) which are guide
fossils of the Lower Ordovician.

The Barrandeocerida are coiled to cyrtocone; their siphuncle is thin-walled
and lacks deposits, orthochoanitic to secondarily cyrtochoanitic. Occurrence:
Ordovician–Devonian. The group includes *Nephriticeras* (= ? *Cyrtoceras*)
(Fig. 96*B*, 4).

The Nautilida embrace the majority of nautiloids from the Devonian to the
present day. They are curved to coiled; the siphuncle is simple, orthochoanitic
and variable in position. There is a large break at the end of the Triassic which
only the Superfamily Nautilaceae survived: *Aturia* (Fig. 96*B*, 5; Palaeocene-
Miocene) with slightly differentiated suture lines. Some of the Palaeozoic and
Triassic Nautilida are strongly sculpted and the whorl section is sometimes
rectangular. Examples: *Germanonautilus* (Triassic), *Trochoceras* (Devonian)
(Fig. 96*B*, 6, 7).

2.   Subclass: Endoceratoidea
    Usually orthocone and holochoanitic (Fig. 95) with thick marginal siphuncle with conical endosiphonal deposits (stacked one inside the other). They are directly derived from the Ellesmerocerida and grew to a length of up to 9 m. Occurrence: Ordovician–Silurian. Examples: *Endoceras* (Ordovician), *Cyrtendoceras* (Ordovician), *Leurocycloceras* (Ordovician–Silurian) (Fig. 97, 1-3).

3.   Subclass: Actinoceratoidea
    Orthocone, thick siphuncle bulging between the septa, endosiphonal canal system and very strong obstruction rings (annulosiphonate). Occurrence: Ordovician–Carboniferous. Examples: *Actinoceras, Armenoceras* (both Ordovician–Lower Silurian) (Fig. 97, 5, 4).

Fig. 97. Endoceratoidea: (1) *Endoceras* (× 0.2; a, b: × 0.25); (2) *Cyrten-doceras* (× 0.25); (3) *Leurocycloceras* (× 0.7). Actinoceratoidea: (4) *Armenoceras* (× 0.3); (5) *Actinoceras* (× 0.3).

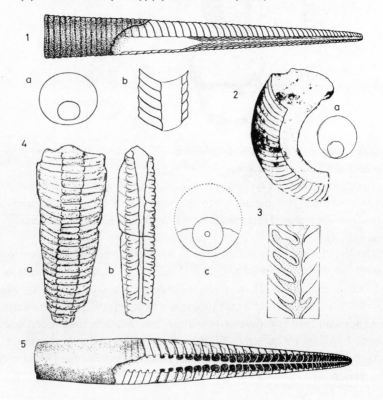

2.   Infraclass: Neocephalopoda
        Comprising the ammonoids and coleoids and their mutual ancestors, the bactritids.

4.   Subclass: Bactritoidea
        The bactritids are externally similar to the Orthocerida, but are small with a thin marginal siphuncle and a siphonal lobe. They can be derived from Ordovician forms with a marginal siphuncle but no siphonal lobe, such as *Bactroceras*. The bactritids embrace the ancestral forms of the ammonites and belemnites and are thus grouped together with them. Occurrence: Ordovician–Permian.
        The oldest form is *Eobactrites* from the Lower Ordovician of Bohemia. It has a circular cross-section and a long narrow siphonal lobe brought about by the marginal siphuncle. The ammonites could well be descended from such forms.
        In *Bactrites* (Fig. 98, 1) (Silurian–Upper Permian) the cross-section is slightly oval, but in *Cyrtobactrites* of the Lower Devonian (Fig. 98, 2) it is well flattened. A lateral lobe is added to the external siphonal lobe.

Fig. 98. Bactritoidea: (1) *Bactrites* (the figure represents a combination, with the left part of the figure showing a lateral view and the right part a ventral view; × 3); (2) *Cyrtobactrites* (× 1).

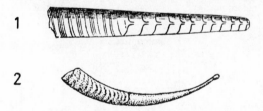

        Parabactrites (Carboniferous–Permian) is brevicone and cyrtochoanitic with narrowly spaced septa. The belemnites can be derived from similar early forms.

5.   Subclass: Ammonoidea
        This subclass includes the goniatites, clymeniids, ceratites and ammonites as such. The term Ammonoidea should be carefully distinguished from the term ammonites which refers to the true ammonites of the Jurassic and Cretaceous. Stratigraphically the Ammonoidea extend from the Lower Devonian to the Upper Cretaceous.
        How the Ammonoidea differ with respect to the Nautiloidea:

   1. the siphuncle is thin, marginal and lacking endosiphonal deposits, siphonal lobe present;

2. the suture is usually complicated;
3. there is predominantly spiral coiling and absence of primitively straight forms.

Differentiation of the suture line

Suture line: line of contact between the septum and the shell wall
Prosuture: suture between the protoconch and the first chamber
Primary suture: suture between the first and second chambers
Lobe: part of the suture which is concave with respect to the aperture
Saddle: part of the suture which is convex with respect to the aperture

External (ventral), lateral and internal (dorsal) lobes appear first in ontogeny and phylogeny; they are thus known as primary lobes. The external lobe (E) lies on the outside of the shell in the centre. The lateral lobe (L) lies on the flank of the shell. The internal lobe (I) lies on the inside of the shell where the two preceding whorls meet (Fig. 99).

Fig. 99. Evolution of the ammonoid primary suture.

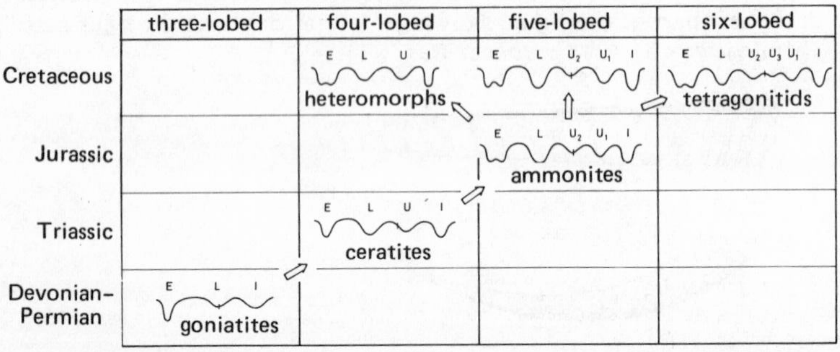

	three-lobed	four-lobed	five-lobed	six-lobed
Cretaceous		E L U I heteromorphs	E L U_2 U_1 I	E L U_2 U_3 U_1 I tetragonitids
Jurassic			E L U_2 U_1 I ammonites	
Triassic		E L U I ceratites		
Devonian–Permian	E L I goniatites			

Lobes can be increased by the division of the internal saddle, which gives rise to umbilical (U) lobes, or by the division of the external saddle which gives rise to adventitious (A) lobes. A saddle arising in the external lobe can lead to the formation of median (M) lobes. Sutural lobes (S-lobes) arise when a saddle is formed in the U-lobe on the umbilical seam followed by subsequent division of lobes and saddles into more or less numerous components.

The complexity of the suture line increases ontogenetically as well as phylogenetically.

Development of the prosuture (Fig. 100): the prosuture has neither internal nor external lobes. It is progressively folded from the Devonian to the Cretaceous. Descriptive terms for the prosuture are asellate, latisellate, angustisellate.

Fig. 100. Evolution of the ammonoid prosuture.

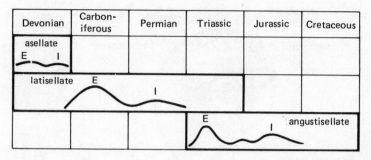

Devonian	Carbon-iferous	Permian	Triassic	Jurassic	Cretaceous
asellate E　　I					
latisellate　　E		I			
			E	I　angustisellate	

Development of the primary suture (Fig. 99): in the Devonian the primary suture consists of internal, lateral and external lobes and the corresponding saddles, and is thus three-lobed (trilobate). In the Permian and Triassic the primary suture is four-lobed (quadrilobate) by the addition of a U-lobe; in the Jurassic and Cretaceous the primary suture is four-, five- or six-lobed (with one to three U-lobes).

The prosuture and primary suture develop quite independently; the former is a larval structure. The terminology presented is genetic and suitable for the recognition and illustration of phylogenetic relationships. It is commonly used by German research workers and, in a modified version, by Russian workers. In the English-speaking areas the morphographic terminology set out in the *Treatise on Invertebrate Paleontology* (Moore & Teichert, 1952ff) is commonly used. It is easier to use for simple (not interpreting) descriptions.

The Ammonoidea originated from the bactritids which themselves are derived from orthocone nautiloids (namely sphaerorthoceratids, themselves derived from michelinoceratids).

Morphographic terms for the Ammonoidea are shown in Figs. 101 and 102.

Fig. 101. Morphographic terms for the elements of the ammonoid suture line.

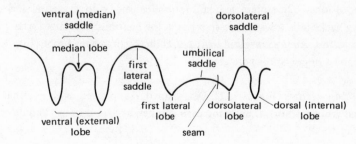

Fig. 102. Morphographic terms and parameters for ammonoids.

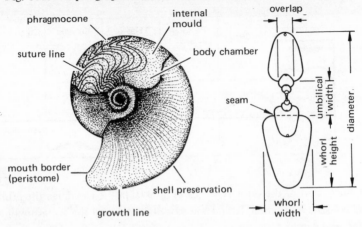

Classification of the Subclass Ammonoidea

1. Order: Anarcestida
2. Order: Clymeniida
3. Order: Goniatitida
} 'goniatites', Palaeoammonoidea suture ± 'goniatitic'

4. Order: Prolecanitida
5. Order: Ceratitida
} 'ceratites', Mesoammonoidea suture ± 'ceratitic'

6. Order: Phylloceratida
7. Order: Lytoceratida
8. Order: Ammonitida
9. Order: Ancyloceratida
} 'ammonites', Neoammonoidea suture ± ammonitic

1.   *Order: Anarcestida*

This order includes the ancestral forms of all later ammonoids. It is distributed throughout the Devonian, and its representatives usually have a small number of lobes. The siphuncle is still retrochoanitic (with the septal necks pointing backwards towards the apex) as in the nautilids. The first forms (*Gyroceratites, Anetoceras*) still had an umbilical gap and no I-lobe (Fig. 103, 1).

The most important genera are:

Agoniatites (Fig. 103, 2): Disc-shaped, compressed, with external marginal groove, whorls rapidly increasing in height. Occurrence: Middle Devonian.

Fig. 103. Order Anarcestida: (1) *Anetoceras* (× 0.5); (2) *Agoniatites* (× 0.2); (3) *Anarcestes* (× 0.5); (4) *Maenioceras* (× 0.75); (5) *Manticoceras* (× 0.25); (6) *Prolobites* (× 0.5); (7) *Beloceras* (× 0.3).

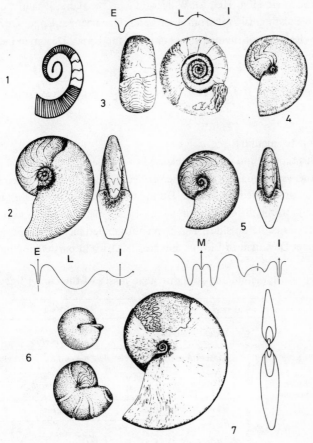

Anarcestes (Fig. 103, 3): Parallel line to agoniatitids, with wider whorls and slower increase in the whorl height. Biconvex growth lines. Occurrence: Lower to Middle Devonian.

Maenioceras (Fig. 103, 4): Flattened flanks; almost parallel external marginal grooves. Occurrence: Upper Middle Devonian (Givetian).

Prolobites (Fig. 103, 6): Narrow umbilicus, straight growth lines, very simple suture initially (in specialised forms such as *Sandbergeroceras* there are many U-lobes). Distributed worldwide. One marked constriction per whorl. Occurrence: Upper Devonian III.

The following dominated in the Lower Upper Devonian:

Manticoceras (Fig. 103, 5): Wide saddle on the flank, E-saddle divided (= M-lobe); this distinguishes it from *Anarcestes*. Occurrence: Upper Devonian I. Explosive development of the manticoceratids in the Upper Devonian I with forms such as:

Beloceras (Fig. 103, 7): Disc-shaped, with largest increase in the number of lobes.

2.   Order: Clymeniida
In this rapidly and highly specialised order the retrochoanitic siphuncle is always internal except in the very early juvenile stages when it is external. The prosuture is latisellate. During the migration of the siphuncle from the outside to the inside, the animal must have maintained its position (ventral side towards the outside). The clymeniids experienced their prime in the Upper Upper Devonian of Europe and North Africa in particular (Fig. 104).

Fig. 104. Clymeniid and goniatite zone fossils of the Upper Devonian.

series and stage				zone fossils
Upper Devonian	Famennian	Wocklumerian	VI	*Wocklumeria Kalloclymenia*
		Dasbergian	V	*Clymenia*
		Hembergian	IV III	*Platyclymenia*
	Frasnian	Nehdenian	II	*Cheiloceras*
		Adorfian	I	*Crickites Manticoceras Pharciceras*

Acanthoclymenia from the Upper Devonian I of New York is regarded as the earliest genus.

Lobe differentiation in the clymeniids: primary lobes E, L, I (*Archoceras* or similar anarcestid as the ancestral form) gave rise to three lines (Fig. 105):

1. The gonioclymeniids represent the conservative stock and the root of the other groups. They have external and internal lobes, a shell that is predominantly disc-shaped, an extremely wide umbilicus, and biconvex growth lines.
Examples:

Gonioclymenia: from the Upper Devonian V (Fig. 106, 1).

Fig. 105. Phylogenetic development of the suture line in the clymeniids.

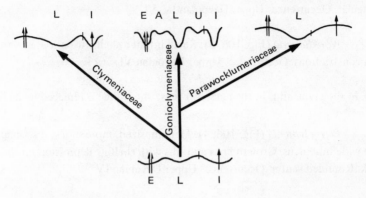

Fig. 106. Order Clymeniida: (1) *Gonioclymenia* (×0.5); (2) *Hexacly-menia* (×0.5); (3) *Wocklumeria* (left: ×0.5; right: ×0.75); (4) *Platy-clymenia* (a: ×0.8; b: ×1.3); (5) *Clymenia* (×0.25); (6) *Parawocklumeria* (×0.65).

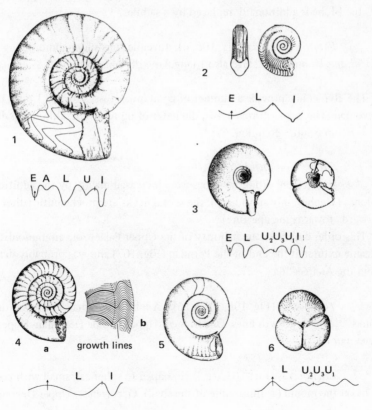

Hexaclymenia (Fig. 106, 2): Small, depressed, whorl-section sub-triangular. Occurrence: Upper Devonian IV–VI.

Wocklumeria (Fig. 106, 3): Small inflated shell. Juvenile stages triangular with constrictions. Occurrence: Upper Devonian VI.

2. In the clymeniids in the narrower sense the E-lobe is replaced by an E-saddle.

Platyclymenia (Fig. 106, 4): Medium-sized, more or less disc-shaped with wide umbilicus. Growth lines and ribs with shallow depression on the outer flank. Rounded venter. Occurrence: Upper Devonian IV.

Clymenia (syn. *Laevigites*) (Fig. 106, 5): Disc-shaped with wide umbilicus; no clear ribbing; lateral lobe wide and rounded. Occurrence: Upper Devonian V–VI.

3. In the parawocklumeriids, which form a small group in the Upper Devonian VI, the I-lobe is additionally replaced by a saddle.

Parawocklumeria (Fig. 106, 6): Juvenile stages homeomorphic with *Wocklumeria*, but adult stage also triangular with constrictions.

The clymeniids provide a number of good index fossils for the Upper Upper Devonian. The largest forms have a diameter of up to 25 cm, although a diameter of 3–10 cm is more common.

3.   Order: *Goniatitida*
    In the Goniatitida the lobes were increased mainly by the addition of A-lobes (U-lobes only subsidiary); the septal necks are prochoanitic (directed forwards towards the aperture).
    This order embraces the majority of the Upper Palaeozoic ammonoids; it became extinct at the end of the Permian (Fig. 107) and was probably derived from the Anarcestida.

Tornoceras (Fig. 108, 1): A large A-lobe and a small L-lobe on the flanks; biconvex growth lines. Occurrence: Upper Middle Devonian–Upper Devonian III.

Cheiloceras (Fig. 108, 2): Disc-shaped to spherical, small with constrictions on the mould (= thickening of the shell). Occurrence: Upper Devonian II.

Fig. 107. Goniatite zone fossils in the Lower Carboniferous.

series and stage				zone fossils
Lower Carboniferous (Dinantian)	Viséan	III	γ β α Aprathian	*Goniatites*
		II	Erdbachian	*Ammonellipsites* *'Pericyclus'*
	Tournaisian	I	Balvian	*Gattendorfia*

Fig. 108. Order Goniatitida: (1) *Tornoceras* ($\times$ 0.3); (2) *Cheiloceras* ($\times$ 0.5); (3) *Sporadoceras* ($\times$ 0.5); (4) *Ammonellipsites* ($\times$ 0.5); (5) *Goniatites* ($\times$ 0.4).

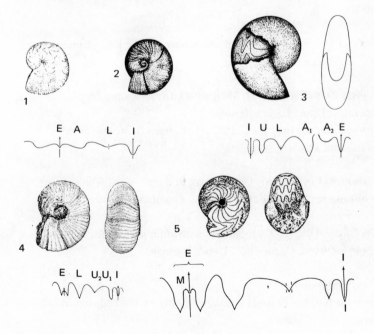

Sporadoceras (Fig. 108, 3): Evolved from *Cheiloceras* but larger; inflated to disc-shaped, with narrow umbilicus (involute). Occurrence: Upper Devonian III.

Gattendorfia: Thick disc-shape; first wide then narrower umbilicus. Occurrence: Lower Carboniferous I.

Ammonellipsites (syn. *Pericyclus*) (Fig. 108, 4): Pronounced continuous ribbing, not interrupted ventrally; with divided E-lobe. Occurrence: Lower Carboniferous II.

Goniatites (syn. *Glyphioceras*) (Fig. 108, 5): Spherical, inflated; with spiral striation. Divided into species according to the number of spiral striations:

Lower Carboniferous III γ	*G. subcircularis*	16	spiral
Lower Carboniferous III β	*G. striatus*	95	striations
Lower Carboniferous III α	*G. crenistria*	75	

In the Upper Carboniferous, for example:

Gastrioceras (Fig. 109, 6): Suture line similar to that of *Goniatites*; whorls wide and depressed; umbilical edge with tubercles. Occurrence: Middle Westphalian.

From the end of the Upper Carboniferous onwards the lobe elements begin to increase.

Schistoceras (Fig. 109, 7): Shell with latticed sculpturing. Guide fossil for the Uppermost Upper Carboniferous.
Eoasianites, so far the only goniatitid to have been found preserved with a jaw apparatus and a radula, is similar.

Perrinites (Fig. 109, 8): Up to 30 cm in diameter, with ammonitic suture line but macrophyllic saddle partitions. Occurrence: Lower Permian.

Cyclolobus (Fig. 109, 9): Saddles with ammonitic frills; the suture line as a whole is curved. Occurrence: Upper Permian.

4.   Order: *Prolecanitida*
Here the number of lobes is increased only by the addition of U-lobes. The siphuncle is retrochoanitic.

Fig. 109. Order Goniatitida: (6) *Gastrioceras* (× 0.5); (7) *Schistoceras* (× 0.75); (8) *Perrinites* (× 0.9); (9) *Cyclolobus* (× 0.6).

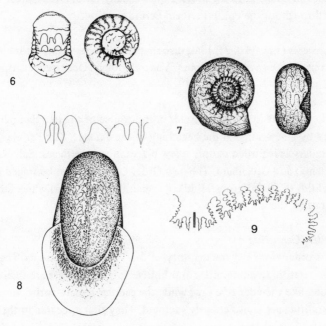

The prolecanitids presumably evolved from the prolobitids towards the end of the Devonian. They represent the ancestral group of all the later ammonoids.

Prolecanites (Fig. 110, 1): Wide umbilicus (evolute); similar to *Psiloceras* in shape. Occurrence: Lower Carboniferous III.

Fig. 110. Order Prolecanitida: (1) *Prolecanites* (× 1); (2) *Medlicottia* (× 0.5); (3) *Sageceras* (× 0.5).

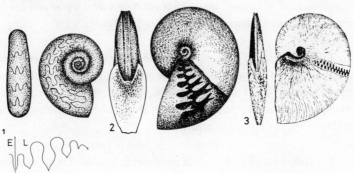

Medlicottia (Fig. 110, 2): Flat, disc-shaped, involute, ventral groove. E-saddle frilled in addition to the lobes. Lobes usually bifid. Occurrence: widespread throughout the entire Permian Period.

Sageceras (Fig. 110, 3): Flat, disc-shaped, all lobes bifid, suture line with numerous simple saddles and lobes. Ventral groove. Occurrence: Alpine Upper Triassic.

The marked decline in the ammonoids in the course of the Permian period was followed by a new surge at the beginning of the Triassic. The Permian/ Triassic boundary is recorded in only a few places in marine facies: Salt Range, Djulfa (Armenia) and Greenland. The new Order Ceratitida can be traced back to the daraelitids (Carboniferous/Middle Permian, Prolecanitida) whose suture line was 'ceratitic' then.

5.   *Order: Ceratitida*
      This order embraces the majority of the Triassic ammonoids. Their suture line is ceratitic to ammonitic. In primitive forms the lobes are finely frilled, looking like the edge of a saw, while the saddles remain entire. In many cases the Ceratitida are conspicuously sculpted. They first appeared in the Upper Permian.

Xenodiscus (Fig. 111, 1): Salt Range, Timor. With almost rectangular cross-section and flat ribs. Occurrence: Upper Permian.

Otoceras (Fig. 111, 2): Triangular cross-section, with a keel and lappets next to the umbilicus. Index fossil for the Lowest Scythian of the Himalayas (lowermost Triassic).

Beneckeia (Fig. 111, 3): Tapering ventral side, involute, lobes usually entire, wide saddles. *B. tenuis*: up to 15 cm in diameter; Uppermost Triassic. *B. buchi*: up to 6 cm in diameter; Lower Muschelkalk.

Ceratites (Fig. 111, 4): Simply forked lateral ribs, and peripheral nodes. In the Upper Germanic Muschelkalk there is a not completely continuous series: it begins with small narrow forms (*atavus-pulcher*), and continues via more conspicuously ribbed forms (*spinosus*) to the uniform series *nodosus-intermedius-dorsoplanus-semipartitus*.

Choristoceras (Fig. 111, 5): First form to be coiled in loose spirals

Fig. 111. Order Ceratitida: (1) *Xenodiscus* (× 0.4); (2) *Otoceras* (× 0.3); (3) *Beneckeia* (× 0.6); (4) *Ceratites* (× 0.4); (5) *Choristoceras* (× 0.6).

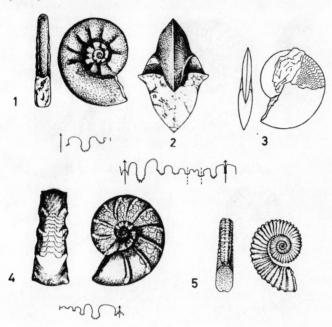

('heteromorph'). The final whorl is detached; pronounced ribbing. Occurrence: Carnian–Rhaetian.

Tropites (Fig. 112, 6): Thick disc to barrel shape; stepped, open umbilicus with tubercles. Occurrence: Carnian–Norian.

Cladiscites (Fig. 112, 7): Thick disc shape; conspicuous spiral striation; suture line strongly frilled with extremely narrow saddle bases. Occurrence: Upper Triassic.

Ptychites (Fig. 112, 8): Thick, discoidal, flat ribs; suture line extensively frilled. *Ptychites* is rare in the Germanic Muschelkalk, more common in the Thetys area, and abundant in some localities of the Arctic. Occurrence: Anisian–Ladinian.

Pinacoceras (Fig. 112, 9): Very narrow discoidal and compressed shell, smooth surface. The most finely divided suture line of all the ceratites.

Fig. 112. Order Ceratitida: (6) *Tropites* (× 0.5); (7) *Cladiscites* (× 0.4); (8) *Ptychites* (× 0.2); (9) *Pinacoceras* (× 0.16).

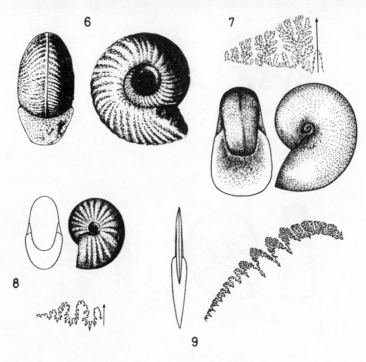

P. metternichi, with a maximum diameter of 1.5 m, is the largest Triassic ammonoid. Occurrence: Upper Triassic.

It was previously thought that there was a particularly sharp break in ammonoid evolution at the Triassic/Jurassic boundary. However, this view will now have to be considerably modified in the light of recent studies. Because of the ease of comparison with older literature we will nevertheless be adhering in general terms to the classification set out in the *Treatise on Invertebrate Paleontology* (Moore & Teichert, 1952ff).

'Ammonites'

Wedekind (1918) contrasted the 'Neoammonoidea' ('ammonites') with the 'Palaeoammonoidea' and 'Mesoammonoidea'. The ammonites differed from the two latter groups mainly by their bipolar division of the suture lines (affecting both lobes and saddles), which in some cases (*Cladiscites, Pinacoceras*) had already been achieved at an earlier stage, and by their stratigraphic status.

According to the traditional point of view, which is also subscribed to in the *Treatise*, the ammonites can be divided into three orders: Phylloceratida, Lytoceratida and Ammonitida. The first of these is known to have existed as early as the beginning of the Triassic, whereas the other two did not appear until the Liassic.

J. Wiedmann (for example, 1973) developed new and different ideas about the phylogeny of the ammonites. In his view, the Lytoceratida and Ammonitida were already present in the Triassic. This is still controversial. Furthermore, Wiedmann was able to demonstrate special differentiations, such as the four-lobed primary suture, in the Cretaceous heteromorphs. This led to the establishment of a fourth order: Ancyloceratida.

6.   *Order: Phylloceratida*

This order was distributed worldwide but uncommon in the Boreal zone.

The characteristic features of the phylloceratids are the phylloid (leaf-shaped) saddle ends and the lituid (crosier-shaped: from the Latin *lituus*, a crook or crosier) internal lobe.

The phylloceratids are thought to have inhabited deeper waters far from the coast. According to recent studies, both the shell resistance and the associated fauna suggest that they preferred depths between 300 and 500 m, comparable with the present-day nautilus with which they may share other ecological similarities.

The order is generally not very diverse and does not show much change; it is thus a conservative line.

Leiophyllites of the Lower Triassic (Lower Scythian–Anisian) is the geologically oldest phylloceratid.

The best-known genus, which is also widespread in Central Europe, is *Phylloceras* (Fig. 113). Long-surviving without great change. Suture with sutural

Fig. 113. Order Phylloceratida: *Phylloceras* (× 0.15).

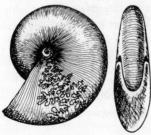

lobe ($U_3 = S$). Suture deeply indented and well branched. I U_1 U_3 (=S), U_4 U_2 E.
Occurrence: Liassic–Upper Cretaceous.

7.   Order: Lytoceratida

This order was also distributed worldwide. It is characterised by a very
finely branched (microphyllic) suture with few but very complex elements. The
internal lobe is bifid with strongly subdivided flanks. In *Lytoceras* (Fig. 114)
the internal lobe shows the extensive division so characteristic of the Lytoceratida.
Occurrence: ? Triassic, Liassic–Upper Cretaceous.

Fig. 114. Order Lytoceratida: *Lytoceras* (× 0.13).

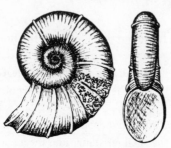

The lytoceratids also preferred deeper waters, although not to the same extent
as the phylloceratids. They must also now be regarded as a fairly unchangeable
conservative line, since the derivation of the Cretaceous heteromorphs, for
example, from the lytoceratids has turned out to be erroneous.

8.   Order: Ancyloceratida

The following genera are assigned to the lytoceratids in the *Treatise*,
but are nowadays combined with the Douvilleiceratida and Deshayesitida into
their own order of the Ancyloceratida. The group of forms described here is
often collectively known as the 'Cretaceous heteromorphs' because its members
diverge from the 'normal' spiral ammonite shape. In contrast to all other
ammonites, the ancyloceratids have a four-lobed suture line.

The heteromorphs were formerly regarded as degenerate forms sentenced to
rapid extinction, but it has since come to light that they were quite a successful
group in ecological and phylogenetic terms (Fig. 115).

They apparently changed their mode of coiling by mutation in a relatively
short period of time, only to return to normal coiling by slow evolutionary steps.
In some cases the return to normal coiling was so complete, e.g. in the case of
the douvilleiceratids, that it is only possible to recognise them as one-time
heteromorphs by their four-lobed primary suture.

Fig. 115. Phylogeny of the Order Ancyloceratida (Cretaceous heteromorphs). Various scales.

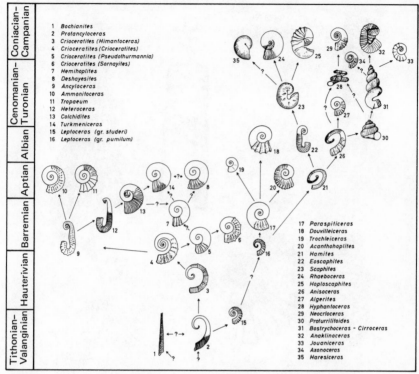

1 *Bochianites*
2 *Protancyloceras*
3 *Crioceratites (Himantoceras)*
4 *Crioceratites (Crioceratites)*
5 *Crioceratites (Pseudothurmannia)*
6 *Crioceratites (Sornayites)*
7 *Hemihoplites*
8 *Deshayesites*
9 *Ancyloceras*
10 *Ammonitoceras*
11 *Tropaeum*
12 *Heteroceras*
13 *Colchidites*
14 *Turkmeniceras*
15 *Leptoceras (gr. studeri)*
16 *Leptoceras (gr. pumilum)*

17 *Paraspiticeras*
18 *Douvilleiceras*
19 *Trochleiceras*
20 *Acanthohoplites*
21 *Hamites*
22 *Eoscaphites*
23 *Scaphites*
24 *Rhaeboceras*
25 *Hoploscaphites*
26 *Anisoceras*
27 *Algerites*
28 *Hyphantoceras*
29 *Neocrioceras*
30 *Proturrilitoides*
31 *Bostrychoceras - Cirroceras*
32 *Anaklinoceras*
33 *Jouaniceras*
34 *Axonoceras*
35 *Haresiceras*

This also disproves the former belief that the heteromorphs were severely
handicapped in their 'struggle for existence' on account of their sometimes
'exotic' shape.

The Ancyloceratida are characterised by a four-lobed primary suture:
I U L E.

Macroscaphites: More or less evolute, final body chamber straight
initially, then hooked. Numerous simple straight ribs. Deeply frilled suture.
Occurrence: throughout the Cretaceous.

Ancyloceras (= *Crioceras*, partly) (Fig. 115, 9): Inner whorls in
a loose spiral, final whorl initially straight, then hooked. Simple radial ribs,
sometimes with nodes. Occurrence: Lower Cretaceous.

Crioceratites (= *Crioceras*, partly) (Fig. 115, 3–6): Coiled like a watch
spring. Oval or almost square cross-section. Ribs predominate over tubercles.
Occurrence: Lower Cretaceous.

Fig. 116. Order Ancyloceratida: (1) and (2) *Baculites* (1a, b: × 0.3,
c: × 0.6; 2: × 0.7); (3) *Turrilites* (× 0.5); (4) *Douvilleiceras* (× 0.5).

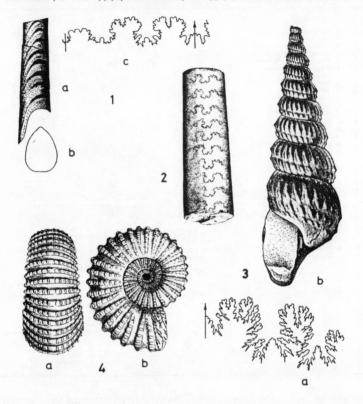

Baculites (Fig. 116, 1-2): Simple rod-like shape, beginning with a spiral
of two whorls (hardly ever preserved). Up to 2 m long in the Maastrichtian. The
baculitids probably lived in association on the sea-bottom. Occurrence: Albian–
Upper Cretaceous.

Turrilites (Fig. 116, 3): Trochispiral, dextrally or sinistrally coiled.
The body chamber occupies two whorls. Occurrence: Albian–Cenomanian.

Bostrychoceras (Fig. 115, 31): Initially like *Turrilites* but with the
final whorl detached and forming a U-shaped hook; simple ribs. Occurrence:
Upper Cretaceous.

Scaphites (Fig. 115, 23): 'Scaphitic' coiling: initially involute, then
detached and then coiled again. Surface with divided and sometimes tubercular
ribs; aperture slightly constricted. Occurrence: Upper Cretaceous.

The so-called false hoplitids, coiled-up descendants of heteromorphs, are also included in the Ancyloceratida, since they also have a four-lobed primary suture and show traces of former uncoiling.

Douvilleiceras (Fig. 116, 4): Ribs interrupted or suppressed by a smooth external band. Ribs disintegrated into rows of tubercles. Occurrence: Lower Middle Albian.

Parahoplites: Ribs with a pronounced curve forming an angle on the outside. Occurrence: Upper Aptian and Lower Albian.

Deshayesites (Fig. 115, 8): Robust ribs projecting forwards on the ventral side, without nodes. Occurrence: Lower Aptian.

9.   *Order: Ammonitida*

This heterogeneous order embraces the Neoammonoidea with the exception of the Phylloceratida, Lytoceratida and Ancyloceratida.

The origin of the first Ammonitida, the psiloceratids, from lytoceratids towards the end of the Triassic was ascertained by Schindewolf in 1962, although in the *Treatise* they are still derived from the phylloceratids.

1.   *Superfamily: Psilocerataceae*

Psiloceras (Fig. 117, 1): Evolute, thin discoidal, smooth or with ribs which do not extend across the external side. Occurrence: Hettangian (Lias $\alpha 1$).

Schlotheimia (Fig. 117, 2): Ribs curved forwards on the ventral side, interrupted along the median line, opposite. Occurrence: *Schlotheimia angulata* in the Hettangian (Lias $\alpha 2$).

Arietites (Fig. 117, 3): Two ventral grooves with a keel in between. Ribs distended into marginal nodes. Occurrence: Hettangian (Lias $\alpha 3$).

The following are derived from the arietitids:

Echioceras (Fig. 117, 8): Evolute, numerous whorls, keel; widely spaced straight ribs. Occurrence: worldwide in the Hettangian and Sinemurian (Lias β). *E. raricostatum* in the Sinemurian (Lias β).

Oxynoticeras (Fig. 117, 4): Flat, disc-shaped, involute. Suture line not very deeply indented; the wide external saddle is divided into two unequal lobes. Occurrence: Lias β. *O. oxynotum* in the Lias $\beta 2$ (Sinemurian).

Fig. 117. Order Ammonitida: (1) *Psiloceras* (× 0.5); (2) *Schlotheimia* (× 0.3); (3) *Arietites* (× 0.5); (4) *Oxynoticeras* (× 0.35); (5) *Amaltheus* (× 0.2); (6) *Pleuroceras* (× 0.3); (7) *Dactylioceras* (× 0.35); (8) *Echioceras* (× 0.5); (9) *Eoderoceras* (× 0.4); (10) *Harpoceras* (× 0.5); (11) *Sonninia* (× 0.3); (12) *Oppelia* (× 0.5).

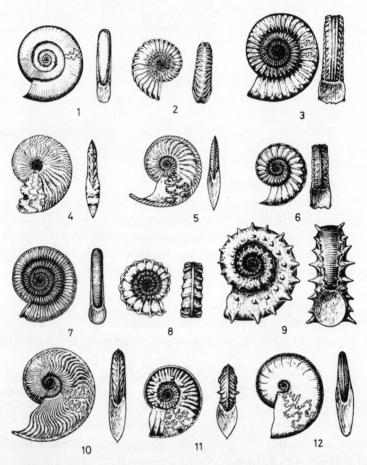

2.   Superfamily: Eoderocerataceae

Ribs widening on the outside, frequently with marginal spines or other thickenings. Evolute, discoidal, flanks often with two rows of nodes connected on the outside by bundles of ribs. *Eoderoceras* from the Upper Sinemurian (Fig. 117, 9).

Androgynoceras ('*Aegoceras*'): Ribs widened externally, not interrupted, no marginal nodes. Occurrence: Lower Pliensbachian (Lias γ).

Fig. 118. Zone ammonites of the Lias.

series and stage			zone fossils
Lias	Toarcian	ƺ ε	*Dumortieria* *Grammoceras* *Haugia* *Hildoceras* *Harpoceras* *Dactylioceras*
	(Domerian) Pliensbachian (Carixian)	δ	*Pleuroceras* *Amaltheus*
		γ	*Prodactylioceras* *Tragophylloceras* *Uptonia*
	(Lotharingian)	β	*Echioceras* *Oxynoticeras* *Asteroceras*
	Sinemurian	α₃	*Euasteroceras* *Arnioceras* *Arietites*
	Hettangian	α₂ α₁	*Schlotheimia* *Alsatites* *Psiloceras*

Amaltheus (*A. margaritatus*) (Fig. 117, 5): The ribs fork at the tapered external side, giving rise to a serrated keel. *Amaltheus* is usually fairly involute. Occurrence: Upper Pliensbachian (Lias δ).

From *Amaltheus* there is a transition to forms with strong ribbing and an almost square whorl section:

Pleuroceras (*spinatum*) (Fig. 117, 6): Occurrence: Upper Pliensbachian (Lias δ).

Dactylioceras (Fig. 117, 7): Evolute, ribs not interrupted on the external side, bifurcated on the outside, a few remaining simple. With double shell which allows for very different states of preservation. Occurrence: Toarcian (Lias ε).

3. *Superfamily: Hildocerataceae (Falciferi)*
The members of this superfamily are usually compressed and furnished with a keel and sickle-shaped (falcate) growth lines and ribs. They often display pronounced sexual dimorphism. Occurrence: Upper Lias–Lower Cretaceous.

Harpoceras (Fig. 117, 10): Weak falcate ribbing; hollow keel with shallow grooves on either side. Subgenus *Eleganticeras* is almost smooth. Occurrence: Toarcian (Lias ε).

The basic blue-print varies considerably at the Lias–Dogger boundary:

Hildoceras: With quadrate whorl section and longitudinal groove on the flanks. Occurrence: Toarcian (Lias ϵ, ζ).

Leioceras: Rather smooth and involute. Occurrence: Lower Aalenian (Dogger α) (*L. opalinum*).

Ludwigia: Stronger keel, more evolute. Occurrence: Upper Aalenian (Dogger β) (*L. murchisonae* among others).

More robust ribbing leads to:

Sonninia (Fig. 117, 11): Keel initially acute, lateral ribs robust, bifurcating nodes at the middle of the flanks. Occurrence: Bajocian.

Oppelia (Fig. 117, 12): Similar to *Leioceras*, but external side wider; ribs more robust with age. Occurrence: Upper Dogger–Lower Cretaceous.

4.   Superfamily: Stephanocerataceae
Forms with regularly ribbed external side. $U_3 = S$. Monoschizotomous, i.e. ribs bifurcating from one point. The following are distinguished according to the differentiation of the sutural lobe: (*a*) Stephanoceratidae with a suspensive ('hanging') sutural lobe; (*b*) Macrocephalitidae with a non-suspensive (straight) sutural lobe.
There is a marked but not fully explored sexual dimorphism, and the microconchs are characterised by the presence of lappets ('ears').

Stephanoceras (Fig. 119, 13): Thick discoidal to spherical. Two to five monoschizotomous bifurcating ribs, tubercles at the middle of the sides. Occurrence: Bajocian (Dogger δ) (*S. coronatum* in the '*coronatum* beds').

Otoites (Fig. 119, 16): A relatively large microconch ammonite with coarse ribbing and large, inwardly curved lappets. Occurrence: Middle Bajocian.

Macrocephalites (Fig. 119, 17): Division of ribs without or with only weak tubercles at the umbilicus. Rather involute. Occurrence: Upper Dogger.

Kosmoceras (Fig. 119, 15): Two rows of tubercles along the external margin, also tubercles on the sides. Macro- and microconchs. Occurrence: Uppermost Dogger.

Fig. 119. Order Ammonitida: (13) *Stephanoceras* (×0.17); (14) *Quen-stedtoceras* (×0.5); (15) *Kosmoceras* (×0.5); (16) *Otoites* (×0.4); (17) *Macrocephalites* (×0.3); (18) *Perisphinctes* (×0.1); (19) *Rasenia* (×0.5); (20) *Gravesia* (×0.25).

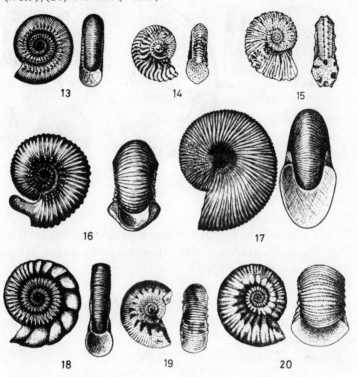

Quenstedtoceras (Fig. 119, 14); A Boreal form whose distribution towards the south reaches almost as far as the Mediterranean. A long ventral projection (rostrum) on the aperture of the microconch. *Quenstedtoceras* displays pronounced sexual dimorphism which has been well documented. Occurrence: Upper Dogger (Callovian).

5.   *Superfamily: Perisphinctaceae*
        Also with regular ribbing on the external side; closely related to the stephanoceratids. Also $U_3 = S$. Sculpturing consists of bifurcating rings developed to different degrees, e.g. monoschizotomous, dischizotomous and polyschizotomous (bifurcating at one, two or several points).

Perisphinctes (Fig. 119, 18): Shell discoidal with ribs bifurcating at the middle of the sides or near the external side. No tubercles, but constrictions present (hence the name); evolute. Occurrence: Malm.

Fig. 120. Order Ammonitida: (21) *Ataxioceras* (× 0.25); (22) *Callizoniceras* (× 0.5); (23) *Tissotia* (× 0.4); (24) *Leymeriella* (× 0.36); (25) *Aulacostephanus* (× 0.5).

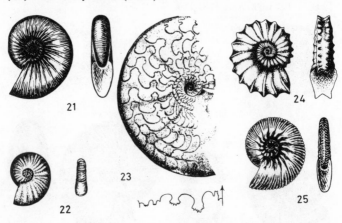

Ataxioceras (Fig. 120, 21): Flattened at the sides with fairly wide umbilicus. Irregular intercalation of ribs (from the Greek *ataxios*, disorderly). Occurrence: Lower Kimmeridgian.

Rasenia (Fig. 119, 19): Widely spaced bulging bifurcating tubercles at the umbilicus, numerous bifurcations, body chamber with ribs. Occurrence: Lower Kimmeridgian.

Gravesia (Fig. 119, 20): Large tubercles giving rise to bifurcations between the umbilicus and the middle of the sides, fading out on the body chamber; shell inflated. Occurrence: Middle Kimmeridgian.

Aulacostephanus (Fig. 120, 25): Similar to *Rasenia*, but with external band at which the ribs terminate. Occurrence: Upper Kimmeridgian.

Virgatites (Fig. 121, 26): Thin discoidal, no tubercles, 'virgatotome' (i.e. several bifurcating points along one rib) ribs extending across the rounded external side. Occurrence: Upper Malm of the Boreal province.

Aspidoceras (Fig. 121, 27): In geologically older forms (*Euaspidoceras*) ribs with tubercles in the juvenile stage, interrupted on the external side; one or two rows of spines on the sides of the adult stage. Occurrence: Malm.

Polyptychites (Fig. 121, 28): Fairly large terminal forms of the perisphinctids, thick discoidal, with tubercles giving rise to bifurcations at the

Fig. 121. Order Ammonitida: (26) *Virgatites* (× 0.25); (27) *Aspidoceras* (× 0.33); (28) *Polyptychites* (× 0.2); (29) *Pachydiscus* (× 0.4); (30) *Hoplites* (× 0.5); (31) *Schloenbachia* (× 0.25); (32) *Flickia* (× 2.5).

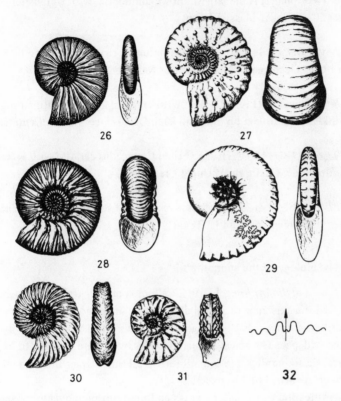

umbilicus; the ribs are diversipartite, i.e. they bifurcate irregularly. Occurrence: Lower Cretaceous (Valanginian).

The two superfamilies Desmocerataceae (6) and Hoplitaceae (7) described in the *Treatise on Invertebrate Paleontology* (Moore & Teichert, 1952ff) require extensive revision. Some of their members (the so-called false hoplitids) are descendants of the 'heteromorphs' and should be assigned to this group. The remaining members are closely related. The following are examples of only some of the relevant genera:

Callizoniceras (Fig. 120, 22): Evolute; ribs or stripes straight or curved forwards; several constrictions. Occurrence: Upper Barremian–Lower Albian.

Hoplites (Fig. 121, 30): A characteristic mid-Cretaceous ammonite with well developed ribs radiating slightly forwards; the ribs are distinctly interrupted externally. Occurrence: Middle Albian (Upper Lower Cretaceous).

Pachydiscus (Fig. 121, 29): Ribs interrupted at the sides; can grow very large. *P. seppenradensis* (= *Parapuzosia seppenradensis*) from the Upper Cretaceous of Münster (Westphalia) is the largest known ammonite, with a diameter of 1.7 m. Occurrence: Campanian–Maastrichtian.

Leymeriella (Fig. 120, 24): Ribs flattened towards the outside, with groove or bulge. Occurrence: Uppermost Lower Albian.

Schloenbachia (Fig. 121, 31): With external keel; ribs bifurcating several times, wrinkled and broken up. Smooth keel. Occurrence: Middle Cenomanian.

Tissotia (Fig. 120, 23): A so-called Cretaceous ceratite with secondarily ceratitic suture line. Occurrence: Upper Cretaceous.

Flickia (Fig. 121, 32): Extreme example of secondary simplification. Occurrence: Cenomanian (Tunisia, Texas).

The biology of the ammonites

1. Sexual dimorphism. In 1962 Callomon and Makowski independently demonstrated the existence of sexual dimorphism in ammonites, which in some cases is very pronounced. Earlier suggestions to this effect were backed by insufficient evidence. The so-called macroconchs, which are thought to be the shells of female ammonites, are sometimes several times larger than the microconchs which are attributed to males.

The identification of the adult stage is an important prerequisite for the recognition of sexual dimorphism. The adult stage is characterised by the approximation of sutures, different sculpturing of the final body chamber or special modifications of the peristome. However, not all ammonoids exhibited a definite end to their growth, and some continued to grow throughout their lives.

In some forms the microconchs are characterised by the presence of apophyses on their apertures, the so-called lappets.

Sexual dimorphism has far-reaching nomenclatorial implications, which have made it a hotly disputed subject.

2. Radula. Radula finds in quite a number of ammonites always revealed the same fairly narrow type of radula with seven very simple teeth for every transverse row, of which the peripheral teeth were several times longer than the rest. From their presence we can conclude that the radula mainly aided the process of swallowing rather than breaking up food.

Given the similarity between the Recent dibranchiates and the fossil ammonites with respect to the architecture of the radula, one can conclude, with some reservations, that there is a close phylogenetic relationship between these two groups. Between the ammonites and the nautilus, on the other hand, there is a larger division which is also suggested by the historical sequence.

3. Aptychi and anaptychi. Aptychi are calcitic structures consisting of two valves and resembling the shells of bivalves. Anaptychi are made up of one unit consisting of a 'horny' (probably originally chitinous) substance and look similar to the two valves of a spread-out aptychus. Since they are occasionally found in the body chambers of ammonoids, they are generally thought to be part of the soft body.

In well-preserved pairs of aptychi it was demonstrated that they share a continuous organic layer on the inside. The calcitic aptychi were thus shown to be nothing more than a more or less thick calcareous layer on the outside of anaptychi.

The one-part anaptychi occur in the Palaeozoic and up to the Cretaceous. From the Toarcian onwards there are also aptychi which consist of two parts and are calcareous (calcitic, in contrast to the aragonitic shell material of the ammonites).

Among the aptychi Trauth (1927–36, 1938) distinguished the following form genera (among others): cornaptychi in the hildoceratids, lamellaptychi in the oppeliids, prestriaptychi in the stephanoceratids, laevaptychi in the aspidoceratids, and rugaptychi in the baculitids.

Interpretation: Because their shape often coincided with the shape of the aperture, both anaptychi and aptychi were generally regarded (following Trauth) as the opercula of those ammonoids associated with them. In the meantime, U. Lehmann has demonstrated that well-preserved aptychi and anaptychi often enclose a 'horny' structure which resembles the upper jaw of Recent cephalopods, so that the two together represent the previously unknown jaw apparatus of the ammonites. Occasionally remains of the radula are found between the two structures.

This finding proved that aptychi and anaptychi acted primarily as lower jaws and were thus part of the jaw apparatus of the ammonoids (Figs. 122 and 123). Some of them, the laevaptychi for example, may possibly have secondarily changed their function to the extent that they may also have acted as a lid.

4. Feeding. The discovery of stomach contents in some ammonites has provided some factual data. The finds include aptychi of smaller ammonites, foraminifera, ostracods, and sea lilies. Analysis of the jaws and radula and a consideration of the mechanism of movement reveal that the ammonites were not hunters but tended to feed on whatever they were able to shovel into

Fig. 122. Model of the jaw apparatus of *Psiloceras*. The lower jaw is an anaptychus.

Fig. 123. Model of the jaw apparatus of *Hildoceras* (*Hildaites*). The lower jaw is covered by a pair of aptychi.

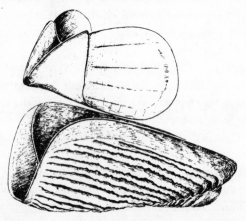

themselves at the sea-bottom by swimming along slowly; plankton cannot be excluded.

5. *Enemies.* Adult ammonoids were hunted and eaten by mosasaurs, turtles, plesiosaurs and other marine reptiles as well as by fishes, crabs, and larger ammonoids. Juveniles and eggs fell prey mainly to fish and other cephalopods. Modern cephalopods represent an important source of food for numerous fish, birds, seals, whales and other cephalopods.

6. *Biotope.* Like the Palcephalopoda, the majority of the ammonites lived in the deeper shelf areas. They were unable to move very fast in the vertical or horizontal directions. Their blunt, sometimes aberrant jaw apparatus (aptychi, anaptychi) suggest that they collected their food rather than hunted it.

6.   Subclass: Coleoidea (Endocochlia)

In the coleoids the external shell is enveloped by a muscular mantle, and the hard parts are surrounded by the soft body. Eight or ten arms furnished with suckers or barbs surround the mouth. Breathing is effected by only one pair of gills. The skin of the Recent forms contains numerous chromatophores. Ink sacs are usually present. The possession of fins and a funnel enables the animals to move and react quickly. In their morphology and efficiency the highly developed eyes are comparable to vertebrate eyes (Fig. 124). The radula has seven to nine teeth in every transverse row.

Fig. 124. Median section of a Recent *Sepia*.

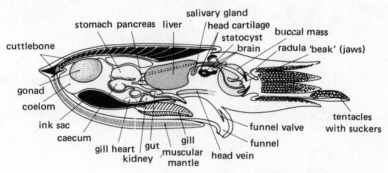

Of the seven coleoid orders distinguished nowadays, the first four are Recent only, with few exceptions, while the other three are known only as fossils.

The ten-armed forms, formerly collectively known as the Decapoda (or better Decabrachia), are now divided into the two orders Sepioidea and Teuthoidea.

1.   Order: Sepioidea (cuttlefish)

The sepioids have ten arms, including two very long ones designed to catch prey, and usually an internal chambered shell. The majority are small to medium-sized, measuring up to half a metre, and live close to the sea-bottom, occasionally burying themselves in loose sand.

This order includes *Sepia officinalis*, well known for its multichambered internal shell, the cuttle-bone (Fig. 125), and the sepia colour of its ink.

Spirula spirula (Fig. 126) is the only dibranchiate species with a spirally coiled shell, although this is enveloped by the soft body (endogastric coiling; Fig. 126). These small animals, measuring only about 5–6 cm, live at depths of 250–1500 m (i.e. at a maximum external pressure of 150 atmospheres). The oldest fossil sepioids described are from the Upper Cretaceous.

Fig. 125. *Sepia* cuttle-bone (× 0.3): (a) ventral view, (b) sagittal section.

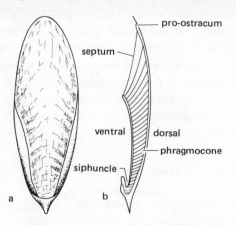

Fig. 126. *Spirula spirula* in its life position; above, the isolated shell (× 0.5).

2.   *Order: Teuthoidea* (squids)

Like the cuttlefish, the squids also have ten arms. This group includes the largest known invertebrates. The shell is reduced to a flat, horny, unchambered 'gladius'. The resulting lack of buoyancy is compensated for by the constant activity of the fins. The squids are continuous and sometimes very fast swimmers, occupying the high seas.

Examples: *Loligo*, the squid, which lives in large swarms; *Architeuthis*, the giant squid, of which the largest complete specimen found was 22 m long with a diameter of 1.5 m.

The Teuthoidea are known since the Lias.

3.   *Order: Vampyromorpha*

The Vampyromorpha represent an eight-armed transitional group between the 'Decapoda' and the octopods. Only a few Recent species are known from the deep sea (1200–4000 m).

4.   *Order: Octopoda* (octopuses)
      The octopods are eight-armed cephalopods without shells or at most with a rudimentary shell consisting of two feeble horny rods. The body is sac-like, the arms are long and very flexible. Members of this group are distributed worldwide; they are predominantly carnivorous bottom-dwellers.
      This group includes *Octopus* and *Argonauta* ('paper nautilus'; the female secretes a brood chamber with two of its arms).

5.   *Order: Aulacocerida*
      In the aulacoceratids, phragmoteuthids and belemnites, which are known only as fossils, the number of arms and gills is not known with certainty. They are assigned to the coleoids because of the similar architecture of their internal skeleton.
      The aulacoceratids are not very diverse as a group; they are primitive in terms of morphology, being closest in line to the Ectocochlia. They have no pro-ostracum (see Belemnitida). Instead, they have a long tube-shaped body chamber, similar to that of the orthoceratids, with a simply curved peristome. The caecum and prosiphon are absent, and the septal necks are prochoanitic. The shape of their 'telum' is similar to that of the belemnite rostrum, but it consists predominantly or totally of an organic substance (conchiolin), while the calcitic layers are reduced. The septa are widely spaced; the phragmocone angle is 5–12°.
      Occurrence: Upper Devonian; Carboniferous–Jurassic, most abundant in the Triassic of the Tethys. There is no close relationship with the belemnites.

      Aulacoceras (Fig. 127): This genus is widespread in the Alpine Triassic. The tela are 10–25 cm long and furnished with about 40 coarse ribs. The phragmocone angle is 5–12°; there is a conspicuous pair of lateral grooves.

      Ausseites (syn. *Atractites*): This genus has a smooth telum and can reach immense sizes (telum 0.5–1.0 m; total length of the animals 4–6 m). Occurrence: Trassic–Lias.

6.   *Order: Phragmoteuthida*
      Of this small group only the genus *Phragmoteuthis* is reasonably well known. It has a three-part fan-shaped pro-ostracum, belemnitic arm hooks and an ink sac. Occurrence: Upper Permian–Upper Triassic of East Greenland and Southern Europe.

Fig. 127. Order Aulacocerida: *Aulacoceras* (× 0.25).

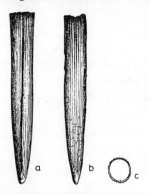

7.   *Order: Belemnitida* (belemnites)

This order embraces the majority of fossil coleoids. The belemnites had a pro-ostracum (corresponding to the dorsal side of the living chamber of orthocone nautilids), a tongue-like projection of the dorsal edge of the shell. The rostrum (guard) consists of calcite. The anterior alveolus accommodated the phragmocone (chambered part) with its narrowly spaced septa and the ventral siphuncle bulging out between the septa. A prosiphon is not known (Fig. 128).

The phragmocone angle is 12-32°. Occasionally specimens are found with an epirostrum (a tube at the tip of the rostrum, which may be 4-30 times the length of the rostrum). The belemnites probably had ten identical short arms and possibly arm hooks (onychites). Occurrence: Lower Carboniferous–Upper Cretaceous.

In general, only the massive rostrum consisting of radiating calcite crystals is preserved. During the lifetime of the belemnite animal the rostrum was probably made considerably lighter than its weight in the fossilised state by the presence of organic substances and cavities. The chambering of the phragmocone, which fitted into the alveolus, provided buoyancy in the centre of the animal and thus created a stable position of equilibrium.

Fig. 128. Schematic longitudinal section of a belemnite shell. AL, apical line; C, chamber; CO, conotheca; P, protoconch; RL, rostral lamellae; S, septum. SI, siphuncle.

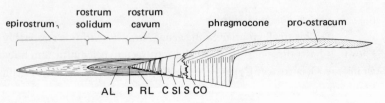

The belemnites were probably nektic animals which lived in swarms close to the surface, predominantly in coastal areas.

Classification is based on the rostrum: cross-section, size, shape, and particularly the variety of grooves, e.g. short grooves at the tip (a few millimetres long), usually fairly flat double lateral grooves, and ventral alveolar grooves. The last are often associated with alveolar slits (interruptions of the rostral layers which look as though they had been carefully sawn through). An important distinguishing feature is the Schatsky value, which is a measure of the shortest distance between the initial chamber and the alveolar margin of the alveolar slit (e.g. in *Belemnella* the distance is up to 4 mm, in *Belemnitella* more than 4 mm) (Fig. 129).

Other characteristics are: vessel impressions which radiate out from the lateral lines, and corrosion forms in which the alveolus is often incomplete in a specific way and looks weathered ('pseudoalveolus'), as in *Gonioteuthis*, for example. The missing part probably consisted of a less preservable (cartilage-like?) material.

The difference between the various forms presumably have no great biological significance. In fact the belemnite animals most probably looked very uniform.

The distribution area of the belemnites was divided into a Boreal and a Tethyan realm. From the Upper Jurassic onwards, belemnites also spread southwards as far as Patagonia, the Antarctic and Australia.

Three suborders are distinguished:

Fig. 129. Longitudinal sections of (a) *Belemnitella*, (b) *Belemnella* (× 0.5).

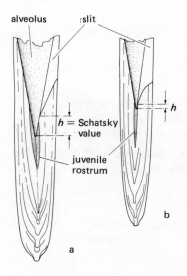

1.   Suborder: Belemnitina

The rostra of the Belemnitina are conical to cylindrical in shape and usually furnished with apical grooves in well preserved specimens. They were most abundant in the Jurassic.

This group also includes the oldest 'belemnites' described from the boundary between the Lower and Upper Carboniferous of the USA. No pro-ostraca have been found, but the rostra themselves have a considerable variety of characteristics.

Examples:

(*a*) Carboniferous belemnites

Paleoconus (Fig. 130, 1): About 3 cm long; one pair of ventral and one pair of dorsolateral grooves.

Fig. 130. Order Belemnitida (Carboniferous belemnites): (1) *Paleoconus* (× 1); (2) *Hematites* (× 0.4).

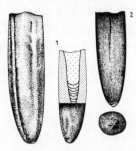

Hematites (Fig. 130, 2): About 6 cm long, no clear lateral grooves, but ventral groove.

(*b*) Jurassic–Cretaceous belemnites
A. 'Dorsolaterales': dorsolateral grooves, but no ventral groove.

Passaloteuthis (Fig. 131, 5): Fairly large belemnite with two dorso-lateral grooves. The alveolus is about two-fifths the length of the rostrum, the apical line is eccentric (formerly known as 'Paxillosi'). Occurrence: Middle to Upper Lias.

Megateuthis (Fig. 131, 1): Very large belemnite with oval cross-section. Dorsolateral grooves are always present, ventrolateral ones usually present. Occurrence: Upper Lias–Upper Dogger.

Fig. 131. Order Belemnitida (Jurassic–Cretaceous belemnites): (1) *Megateuthis* (× 0.07); (2) *Odontobelus* (× 1); (3) *Dactyloteuthis* (× 0.25); (4) *Hastites* (× 0.5); (5) *Passaloteuthis* (× 0.33).

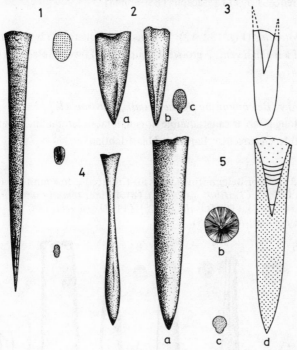

B. 'Tripartiti': paired dorsolateral grooves, ventral groove not paired.

Odontobelus (Fig. 131, 2): This form differs from the 'Paxillosi' by the presence of a ventral apical groove. The alveolus is very eccentric and curved, the phragmocone angle is wide (23–30°). Occurrence: Upper Lias–Lower Dogger.

Dactyloteuthis (Fig. 131, 3): Laterally compressed with flat ventral groove. Occurrence: Upper Lias.

C. 'Hastati': No apical grooves; ventral alveolar canal (can be derived from the ventral apical groove of the 'Tripartiti') and alveolar incision. Occurrence: Dogger–Upper Cretaceous.

Hastites (Fig. 131, 4): Club-shaped; the centre of gravity is shifted towards the end of the club (in contrast to *Hibolites*). The rostrum is very long and slender without grooves. The large phragmocone is rarely preserved. Occurrence: Lias–Middle Dogger.

Oxyteuthis (Fig. 132, 11): The rostrum is large and often more or less flattened on the ventral side. There is no alveolar depression. Double lateral lines present. Occurrence: Lower Cretaceous (Barremian).

Aulacoteuthis (Fig. 132, 12): This genus differs from *Oxyteuthis* by the presence of a median ventral groove. Occurrence: Lower Cretaceous (Middle Barremian).

2.   *Suborder: Belemnopsina ('Canaliculati', 'Gastrocoeli')*
    With long alveolar canal; double dorsolateral or lateral lines. Derived from the hastitids. Occurrence: Bajocian–Maastrichtian.

Fig. 132. Order Belemnitida (Jurassic–Cretaceous belemnites): (6) *Belemnopsis* (× 0.5); (7) *Hibolites* (× 0.3); (8) *Actinocamax* (× 0.5); (9) *Gonioteuthis* (× 0.75); (10) *Duvalia* (× 0.5); (11) *Oxyteuthis* (× 0.7); (12) *Aulacoteuthis* (× 0.8).

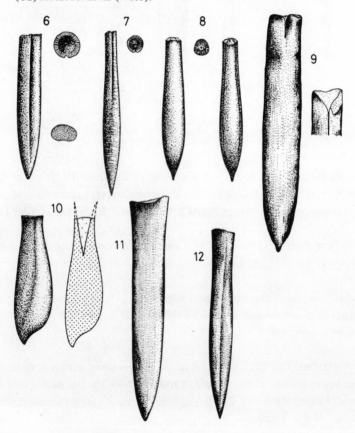

Belemnopsis (Fig. 132, 6): The alveolar canal almost extends to the tip, with alveolar slit. Occurrence: Middle-Upper Dogger.

Hibolites (Fig. 132, 7): Marked spear to club shape, well developed lateral lines; a ventral groove extends well backwards. Occurrence: Middle Dogger-Upper Malm.

Neohibolites: Small rostrum with alveolus, ventral groove restricted to alveolar part of rostrum. Occurrence: Aptian-Cenomanian.

Gonioteuthis (Fig. 132, 9): Well developed pseudoalveolus; rostrum more or less cylindrical. Occurrence: Cenomanian-Campanian.

Actinocamax (Fig. 132, 8): Similar to *Gonioteuthis*, but more club-shaped.

Belemnitella (Fig. 129a): The large rostra of this genus have lateral lines and branched vessels and often granulations at the sides. The ventral margin of the alveolus is deeply incised. The alveolar angle measures 19-26°. The blunt end is furnished with a point (mucro). The Schatsky value exceeds 4 mm. Occurrence: Santonian-Maastrichtian.

Belemnella (Fig. 129b): Similar to *Belemnitella*, but the alveolar angle is 12-21° and the alveolar slit emanates from the level of the protoconch (i.e. Schatsky value of 0-4 mm). Occurrence: Maastrichtian.

(The Belemnoteuthida, e.g. *Belemnoteuthis*, *Diplobelus*, and others, probably also belong to this suborder.)

3.   *Suborder: Duvaliina ('Nothocoeli')*
Short, laterally compressed rostrum; alveolar groove present only on the dorsal side. Occurrence: Upper Dogger-Cretaceous. Examples: *Duvalia* (Fig. 132, 10), from the Neocomian of the Tethys.

The youngest belemnite-like forms, the so-called neobelemnites, have a very much reduced rostrum, while the phragmocone has a tendency towards coiling up. They are sometimes assigned to the Sepioidea. Examples: *Bayanoteuthis* and *Vasseuria* from the Eocene.

7.   Class: Coniconchia (syn. Cricoconarida)
Small, extinct, pointed conical and ringed calcareous shells, usually

measuring only a few millimetres, with an apical angle of between 2° and 8°, are collectively known as the Coniconchia. The initial part of the shell is chambered, but has no siphuncle. The aperture is smooth and entire. In certain Palaeozoic rocks, particularly those of the Upper Devonian, forms of the Order Dacryoconarida occur in virtually rock-forming numbers. Because of their relatively rapid rate of evolution, some of them are good index fossils (Figs. 133 and 134).

Fig. 133. Schematic representation of a representative of the Tentaculitida (×10).

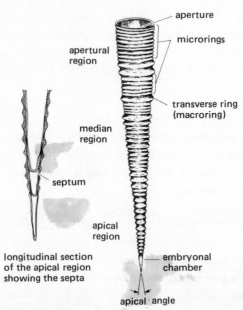

The animals probably had either a pelagic-planktonic (dacryoconarids) or a nekto-benthic lifestyle (tentaculitids).

The zoological status of these microfossils has still not been finally settled. They were formerly assigned to a host of different classes. The composition of the shell material, the structure of the shell and the chambering all seem to suggest that they are members of the mollusc phylum.

Occurrence: Ordovician–Upper Devonian.

1.   Order: Tentaculitida

Shells 15–30 mm long; small apical angle; simple point; with transverse rings; with septa. Example: *Tentaculites* of the Lower Devonian (Fig. 133).

Fig. 134. Schematic representation of a representative of the Dacryoconarida (×15).

Ultrastructure of the shell material

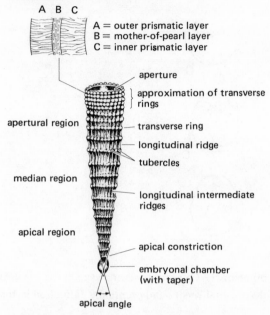

A = outer prismatic layer
B = mother-of-pearl layer
C = inner prismatic layer

aperture

approximation of transverse rings

apertural region

transverse ring

longitudinal ridge

tubercles

median region

longitudinal intermediate ridges

apical region

apical constriction

embryonal chamber (with taper)

apical angle

2.   *Order: Dacryoconarida*

Smaller than the tentaculitids, with a larger apical angle; with broad
ripple-like rings; initial part inflated into a 'tear-drop', no chambers or longitudinal
ribbing. Examples: *Nowakia* (Lower to basal Upper Devonian), *Styliolina*
(Devonian) (Fig. 134).

8.   Class: Calyptoptomatida

The extinct class of the Calyptoptomatida includes conical calcareous
shells which usually have a roughly triangular or occasionally circular to oval
cross-section. With an apical angle between 10° and 40° the length of the shell
is very variable, ranging between 1 and 50 mm. The chambering of the initial
part observed in some forms is an important feature (Fig. 135).

The aperture could be closed off by an operculum. Occasionally two leaf-
shaped curved calcareous lamellae project from the aperture; these are thought
to be fin supports.

Three orders are distinguished according to the shape of the apical part:

1. Order: Hyolithida (conical apical region)
2. Order: Globorilida (spherical apical region)
3. Order: Camerothecida (cylindrical chambered apical region)

Fig. 135. Class Calyptoptomatida: *Hyolithes.* (a) Lateral view,
(b) transverse section, (c) operculum (inside), (d) operculum (outside).
Approximately natural size.

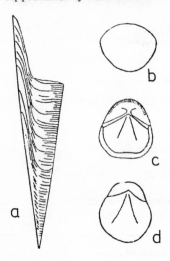

The beginning and prime of the Calyptoptomatida were in the Cambrian,
followed by a gradual decline in diversity and eventually extinction in the
Permian.

The zoological status of the Calyptoptomatida has long been a bone of
contention, and they have variously been regarded as worms, pteropods,
cephalopods or as problematical forms. However, the composition of the shell
material and the occasional chambering suggest that they were probably molluscs.

10. Phylum: *Aschelminthes* (Nemathelminthes, round worms)

The Aschelminthes are simple unsegmented worms with no blood vessels
or coelom. The phylum includes the hair worms (Nematomorpha), of which the
first fossil representative, *Gordius tenuifibrosus*, is preserved with its soft parts in
the Eocene of the Geisel valley (Halle/Saale).

11. Phylum: *Annelida* (segmented worms)

The annelid body is metameric and consists of numerous segments.
The mouth opening is ventral, and there is often a special jaw apparatus. The
annelid fossil record consists of tracks, impressions made by the soft body,
special burrows, and elements of the jaw apparatus. They have been shown to
have existed since the Precambrian.

Dickinsonia costata Sprigg and *Spriggina floundersi* Glaessner from the Ediacara fauna of South Australia (about 670 million years old) are thought to be the oldest annelid worms (138g, h).

1.   Class: Polychaeta
The polychaetes are clearly segmented, predominantly marine annelids. On the sides of the segments, which are furnished with bristles, there are parapodia, usually consisting of two branches. Most polychaetes are either part of the mobile benthos or sessile, with many spending some or all of their time in tubes consisting either of secreted calcareous material or cemented foreign bodies. Tube-dwelling polychaetes, e.g. *Sabellaria* and *Serpula*, often live in enormous assemblages forming reef-like complexes. Polychaetes are known to have existed since the Precambrian (Ediacara).

1.   Order: Errantia
Polychaetes with a well defined head region and usually well developed jaw apparatus. They are mostly part of the vagile benthos. Particularly well preserved fossil forms are found in the Burgess Shale (Middle Cambrian) of British Columbia. The impressions of the soft body reveal the segmentation, bristles, bundles of bristles, tentacles and parts of the alimentary canal. The lugworm *Arenicola marina*, which is particularly common in the North Sea mud-flats along the North Sea coasts where it lives in U-shaped burrows, is a characteristic Recent species. *Arenicola* burrows can be recognised at the surface by the worms casts and by the little pits in the sand.

2.   Order: Sedentaria
Usually sessile polychaetes without a jaw apparatus, which mostly live in solid tubes. Members of the Family Serpulidae which live in calcareous tubes are known in fossil form. The tubes may be closed by a lid; fossil serpulid lids are rarely found. The surface sculpturing of the tubes may vary considerably (Fig. 136).
A comparison of serpulid tubes with some very similar gastropod shells shows that both may be chambered, but that the composition of the shells is different (Table 6 and Fig. 137).
The classification of Recent serpulids is mainly based on the characteristics of the soft body, the ring of tentacles and the bristles, while that of fossil forms relies solely on characteristics of the tubes such as the shape of the cross-section and the sculpture.
Examples:

Fig. 136. Order Sedentaria: Recent serpulid with an operculum inside its tube (× 2).

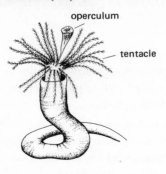

Fig. 137. Shell composition of (a) a serpulid, (b) a gastropod.

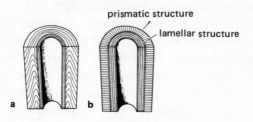

Table 6. *A comparison of the shell composition in serpulids and gastropods*

Serpulidae	Gastropoda
Tube cross-section	
Outer and inner layers made up of concentric deposits	Outer layer consisting of radiating fibres; inner layer made up of concentric deposits
Longitudinal section	
Outer layer with parabolic structure curved towards the oral side; thin inner layer laid down in fine layers parallel to the inner surface	Outer layer made up of vertical fibres; inner layer of thin fine deposits parallel to the inner surface

Serpula (Fig. 138, c): Tubes irregular, often circular and encrusted. The generic name *Serpula* is used in a very wide sense as far as fossil material is concerned. In the Upper Malm of Northern Germany the genus sometimes acts as a rock-former in a biostrome-like fashion, so that the sequence in question is known as a serpulite. The most common species in the serpulite is *Glomerula gordialis.* Occurrence of *Serpula*: Silurian–Recent.

Spirorbis (Fig. 138, a, b): Planispirally or trochospirally coiled tubes forming a snail-like shell. The tubes are fixed by their apex, while the last whorl is usually free. The same species will often have dextrally and sinistrally coiled shells. Many species in the Upper Cretaceous. Occurrence: Ordovician–Recent.

Rotularia (Fig. 138, e): Shell usually coiled in a flat spiral, posterior end of the tube usually attached to the substrate; initial part of the tube coiled up irregularly. Genus measures from a few millimetres to several centimetres. Occurrence: Cretaceous–Eocene.

Fig. 138. Order Sedentaria: (a) and (b) *Spirorbis* (× 2); (c) *Serpula* (× 2); (d) *Ditrupa* (× 3); (e) *Rotularia* (× 1); (f) *Cornulites* (× 1); (g) *Dickinsonia* (× 0.4); (h) *Spriggina* (× 1).

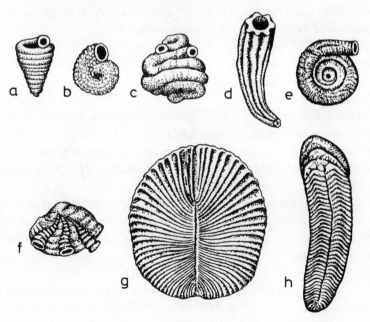

Fig. 139. Above, a longitudinal section of a polychaete head (*Eunice punctata*; × 8). Below (a, b), various scolecodonts (× 15) from the Zechstein limestone (Upper Permian).

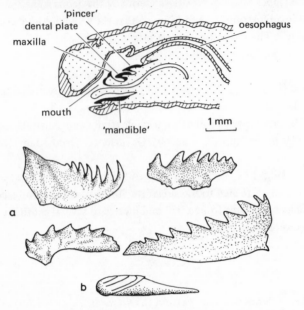

Ditrupa (Fig. 138, d): Tubes not fixed, cylindrical to club-shaped and open at both ends. The slightly curved tubes, about 1 cm in length, are similar to scaphopod shells. Occurrence: Cretaceous–Recent.

Cornulites (Fig. 138, f): Conical tube with well defined regular sculpture. Often grows on brachiopod valves. Occurrence: Ordovician–Devonian.

The serpulids sometimes aid stratigraphic conclusions. As far as the sculpture of the tubes is concerned, it is remarkable that forms occurring from the Ordovician to the Triassic are almost exclusively smooth. Only from the Jurassic onwards does the sculpturing become more differentiated, reaching a climax in the Upper Cretaceous and in the Tertiary.

The two following genera together make up a considerable part of the fossil assemblage at the Precambrian site of Ediacara (Southern Australia). Although their affinities cannot be ascertained, they are generally thought to belong to the annelid worms.

Dickinsonia (Fig. 138, g): The largest of several different species grew up to a length of almost 40 cm. It is oval in shape with numerous transverse ridges (between 20 and several hundred). There are no visible traces of eyes or legs. Its shape is similar to that of the Recent annelid worm *Spinther* which parasitises sponges.

Spriggina (Fig. 138, h): A narrow flexible body up to 4 cm in length and a horseshoe-shaped head shield. The body carried up to 40 pairs of lateral projections (parapodia) with delicate spines on the ends. *Spriggina* is similar in shape to the Recent marine worm *Tomopteris*. The head shield is reminiscent of the head shields of certain trilobites, but it does not seem probable that trilobites are derived from *Spriggina*-like forms.

Scolecodonta

The chitinous or partly calcified jaw segments of fossil bristle worms are known as scolecodonts. In Recent free-living annelids the pharynx can be turned inside out like a pocket. The various jaw elements are arranged in the pharynx in such a fashion that prey can be grasped. The following elements are distinguished: paired mandibles, maxillae, pincers, dental plates and an unpaired serrated plate (Fig. 139). When moulting, the annelids also discard the jaw apparatus. The size of the individual elements varies between about 0.1 and 5 mm. As in the case of the Conodonta, the jaw elements are found in isolation predominantly in dark shales and limestones. Fossil scolecodonts are known since the Cambrian; they are particularly abundant from the Ordovician to the Devonian.

12. Phylum: Arthropoda (jointed-legged animals)

The arthropods are by far the most successful animal phylum. With close to one million species they alone represent 75% of all known present-day species. Ninety per cent of present-day arthropod species belong to the Class Insecta. At the same time, the arthropods make up one of the oldest phyla which was already in existence at the beginning of the Cambrian with highly evolved forms.

The arthropods were very successful in their conquest of terrestrial biotopes, as exemplified by the Arachnida (spiders) which were represented as early as the Silurian by the orders Scorpionida and Myriapoda (millipedes). Among the fossil insects the orders Palaeodictyoptera and Megasecoptera (dragonflies in the wider sense) in particular are known from the Carboniferous. The genus *Meganeura* with a maximum wing-span of about 70 cm is particularly remarkable.

When land was colonised in the Silurian, the aquatic ancestors of the arthropods were already in possession of a number of anatomical prerequisites for terrestrial life: e.g. they had sufficient protection from evaporation, a differentiated nervous system and functional legs. It is possible that only the tracheae were newly acquired by the terrestrial arthropods. The relative inefficiency of the tracheae places severe limitations on size.

A very significant step in arthropod evolution was the acquisition of an exoskeleton, which was present in many marine arthropods as early as the Cambrian. The exoskeleton or cuticle is secreted by the epidermis of the animal as a hard casing (integument). The cuticle consists mainly of chitin and is frequently hardened (sclerotised) by impregnation with calcium carbonate or calcium phosphate. However, in most cases only the (outer) exocuticle is impregnated, while the (inner) endocuticle remains soft. An outer epicuticular coating consisting of a waxy substance prevents the penetration of water and acids. Flexible hairs (setae) can penetrate the cuticle from the epidermis. Infoldings of the sclerotised cuticle of various shapes give rise to apodemes, the attachment sites for muscles.

The arthropod cuticle always adheres strictly to the segmentation of the body. In the typical case, each segment is protected by a dorsal plate (tergite) and a ventral plate (sternite). Lateral sclerotised plates (pleurites) are present in only a few arthropods (e.g. trilobites). All arthropods characteristically moult several times during their lives at more or less regular intervals (ecdysis). Since the rather inflexible sclerotised cuticle cannot grow along with the animal, it is shed from time to time. During this process the endocuticle is dissolved so that the exocuticle becomes detached and peels off along a non-sclerotised moulting suture. The course of these sutures is often of taxonomic importance, e.g. in the trilobites. Not only is the exocuticle renewed during moulting, but also all internal organs lined with chitin, such as the fore-gut and the rectum, parts of the sexual organs and the tracheae. The discarded parts of the cuticle are known as exuviae.

The body is usually divided into several regions: the head (cephalon), thorax (head and thorax are often fused into a cephalothorax or prosoma), and abdomen or pygidium. In the head region the paired appendages of the arthropods are extensively modified for the purposes of recognising and taking up food, looking after broods, etc. The paired appendages in the thoracic region are normally developed as walking legs for locomotion (Fig. 140). They consist of jointed segments with complicated homological relationships. The basic elements are the coxa ('hip'), femur ('thigh'), tibia ('lower leg') and tarsus ('foot'), but these may be further subdivided.

The individual appendages may have rigid or mobile lateral processes; those pointing outwards are called exites, those pointing inwards endites. The exites of the first three appendages of crustaceans are known as pre-epipodite, epipodite and exopodite. The first two are usually developed as gills. Crustaceans have

Fig. 140. Terminology for crustacean appendages.

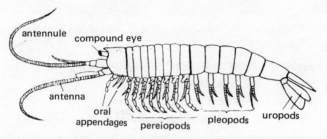

two-branched appendages consisting of the outer exopodite and the inner endopodite (walking leg) which both emanate from the basipodite. Fusion of the basipodite, the coxa and precoxa may give rise to the so-called sympodite. In the trilobites, the branch used as a walking leg is known as the telopodite. In the case of the trilobites, a multi-membered lateral branch (pre-epipodite) is situated quite high up the base of a sympodite. It is uncertain whether the former is an independent precoxa or whether it is homologous with the basal portion of a large coxa. The trilobite 'pre-epipodite' may be homologous with the crustacean exopodite, epipodite or pre-epipodite (Fig. 141). The evaluation of the relationship between the trilobites and the crustaceans is strongly dependent on the interpretation of these homological relationships.

The arthropods represent a very old phylum which specialised very early on. Although unequivocal fossil evidence is still lacking, there is no doubt, on the basis of comparative anatomical studies, that the Recent annelids (segmented worms) and the arthropods shared a common ancestor. The arthropod body plan merely seems to represent a progression from the annelid body plan. The most

Fig. 141. A comparison of trilobite and crustacean appendages.

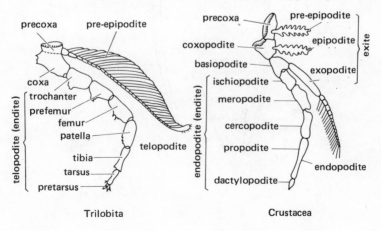

important characteristic of both phyla is the complete internal and external segmentation.

The following summary includes only some of the subphyla of fossil importance:

1. Subphylum: Trilobitomorpha (Cambrian–Permian)
2. Subphylum: Chelicerata (Cambrian–Recent)
3. Subphylum: Crustacea (Cambrian–Recent)
4. Subphylum: Insecta (Carboniferous–Recent)

The subphyla: Pycnogonida (sea spiders, Devonian–Recent), Onychophora (Cambrian–Recent), Myriapoda (millipedes, Carboniferous–Recent), Pentastomida (Recent) and Tardigrada (water-bears, Recent) will not be discussed here because they are of no great palaeontological significance.

Our classification of the arthropods also follows that of the *Treatise on Invertebrate Paleontology* (Moore & Teichert, 1952ff). However, it should be borne in mind that doubts, in some cases serious ones, have been voiced about the classification; but in our view, as far as the beginner is concerned, the practical value of the comprehensive and detailed classificatory system set out in the *Treatise* still outweighs these doubts.

As an example of more recent phylogenetic concepts we have included a slightly modified version of an evolutionary tree relating particularly to the Arachnata (here Arachnata = Trilobitomorpha + Chelicerata), in which the special status of the Olenellida (see p. 197) is taken into account (recognised by the worm-shaped posterior part of the body = telosome (opisthothorax), the absence of the

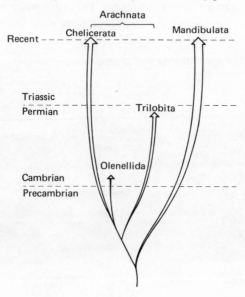

pygidium and the hypoparian facial suture). The original version of this evolutionary tree is published in the works of K. S. Lauterbach (1980). In this particular scheme the Olenellida remain close to the Chelicerata and are clearly separated from the 'eutrilobites'.

1. Subphylum: Trilobitomorpha

The Trilobitomorpha embrace the primitive arthropods with 'trilobitomorphic' appendages. These are characterised by the presence of a lateral branch (the pre-epipodite) which emanates from the base of the walking leg or telopodite. In addition to the trilobites, this group includes a number of small forms from the Middle Cambrian Burgess Shale of British Columbia, which strongly resemble other arthropods in their appearance. The Merostomoidea (main representative: *Sidneya*) are thus similar to the Merostomata, while the Pseudonotostraca (e.g. *Burgessia*) resemble some crustaceans. The Marrellomorpha, which are represented by a single genus *Marrella*, differ from the trilobites by their long genal spines. The genus *Cheloniellon* from the Lower Devonian Hunsrück Shale is the only younger form to be assigned to the Trilobitomorpha.

1. Class: Trilobita

Trilobites are extinct marine arthropods of the Palaeozoic Era. Thanks to their great diversity and abundance of individuals, their relatively easy identification and rapid evolution, they are able to provide us with numerous guide fossils. They existed throughout the Palaeozoic from the Lowest Cambrian. About 1300 genera have been described.

The animals were protected by an integument on their dorsal side which continued ventrally along the margin as a narrow flange (doublure). Most of the ventral side was not covered by an integument. The dorsal integument is divided into three regions from front to rear; the cephalon = head shield, the thorax = body, and the pygidium = tail shield (Fig. 142). There are also three vertical divisions: pleura (left), axis and pleura (right).

The dorsal integument consists of a two-layered calcitised cuticle with a thin external prismatic layer and a thick stratified inner layer. The integument was periodically discarded by moulting (ecdysis) in order to keep pace with the increasing size of the animal.

Both in the cephalon and pygidium, the original skeletal elements have fused together. Only in the thoracic region, which consisted of 2–40 segments, was it possible to move the segments with respect to each other; metameric somites.

Fig. 142. Dorsal view of a trilobite integument (schematic).

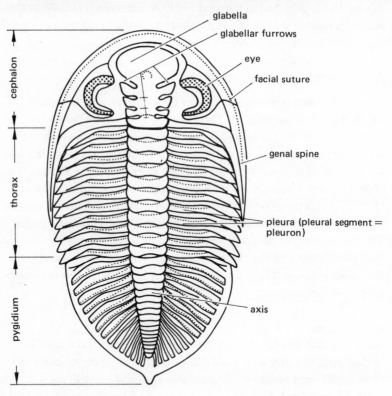

The cephalon has a raised central hump, the glabella, which may be indented by furrows (traces of original segmentation). So-called facial sutures separate the fixigenae from the librigenae which are discarded during moulting. The occipital ring which backs on to the glabella forms the transition to the body. The glabella, fixigenae and the occipital ring make up the cranidium (Fig. 143).

Fig. 143. Characteristic features of a trilobite cephalon.

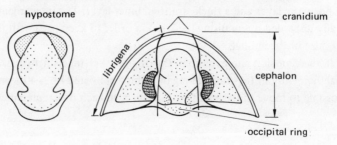

The facial sutures facilitated moulting. A variety of cephalic sutures are distinguished, depending on the course they take: e.g. proparian, opisthoparian, gonatoparian, metaparian, and protoparian or hypoparian (Fig. 144). The facial suture runs over the palpebral lobe along the inside of the compound eyes. The palpebral lobes may be connected to the glabella by the ocular ridges.

The visual area of the compound eyes is furnished with a variable number of lenses (maximum 15 000), under each of which there is a crystalline cone. Trilobites with no eyes at all also existed. When the individual polygonal lenses touch each other along their edges, the arrangement is known as holochroal. If the lenses are hemispherical and separated from one another, the eye is known as a schizochroal eye. In contrast to the chitinous eyes of Recent arthropods, the lenses of trilobite eyes consist of calcite crystals in the correct optical orientation, i.e. the C-axis is at right angles to the surface.

On the ventral side of the cephalon there may be three plates: the rostral plate; the taxonomically important, more or less egg-shaped hypostome, which lay in front of the mouth and behind the rostral plate; and the very small meta-

Fig. 144. Various types of facial sutures in the trilobites.

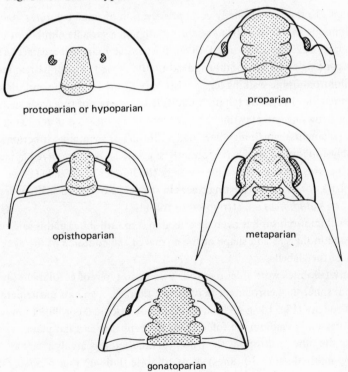

protoparian or hypoparian

proparian

opisthoparian

metaparian

gonatoparian

stoma, which lay behind the mouth opening and has only been described in a few genera (Fig. 145).

The thorax consists of a number of articulating segments. Each segment consists of an axial ring and two lateral pleura (singular pleuron). Typically, each pleuron is indented by an oblique depression, the pleural furrow. Together the axial rings form the axis.

The pygidium consists of a variable number of fused segments. Like the cephalon, it is articulated with the thorax. The pygidium can be very small or quite large, surpassing the cephalon in size.

Fig. 145. Cephalon of *Paradoxides*: (a) dorsal view, (b) ventral view showing the rostral plate (vertical lines) and hypostome.

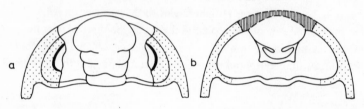

Thanks to the exceptionally good preservation of some trilobites, one pair of uniramous antennae in front of the mouth and a variable number of biramous appendages have been identified. In contrast to the branched crustacean appendage which consists of an endopodite and exopodite as well as a gill branch (epipodite), the trilobitomorphic walking leg consists only of the telopodite and the pre-epipodite (Fig. 141). The telopodite fulfilled a number of functions such as digging, browsing and crawling. The pre-epipodite consisted of relatively rigid spines (filaments) which may have had a filtering or swimming function, but probably functioned as gills (as formerly suggested). Well preserved appendages and the corresponding tracks are found in the Bundenbach Shale (Lower Devonian), for example. Resting traces in the shape of paired shallow depressions (*Cruziana = Bilobites*) are attributed to trilobites.

Sexual dimorphism has been described in some trilobites and is said to be expressed in the size and shape of the animal or the contours of the eyes or the height of the glabella.

Many trilobites were able to enroll. Two main types of enrollment are distinguished: spheroidal enrollment in which all thoracic segments participate more or less equally (Fig. 146), and the less common discoidal enrollment in which the thorax and pygidium are folded over the cephalon as a flat plate.

Like the other arthropods the trilobites moulted repeatedly (they are thought to have moulted about 30 times). Most trilobite finds are thus probably discarded

Fig. 146. Schematic representation of (spheroidal) enrollment in a phacopid.

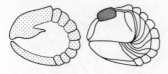

integuments (exuviae) rather than remains of carcasses. One interesting moulting position is the so-called Salterian moulting described in the Devonian Phacopida. The thorax and pygidium are embedded with the concave side pointing downwards, while the cephalon is embedded with the concave side pointing upwards, i.e. rotated by 180° (Fig. 147). For obvious reasons, this type of embedding has only been found in fine clastic sediments (calm sedimentary conditions).

Fig. 147. Salterian moulting as seen in a phacopid (*Trimerocephalus*, × 0.5).

In the trilobites, three different ontogenetic stages are distinguished. The earliest, protaspis stage consists of a uniform but already segmented plate; the length of the larva is $\frac{1}{4}$-1 mm. The meraspis stage begins with the first appearance of a joint between the cephalon and pygidium. The larvae are 6 to 12 times the size of the largest protaspid stage. At the holaspis stage the final number of segments is laid down. Subsequent moults allowed for a considerable increase in size. Most adult trilobites grew to a size of between 3 and 8 cm. The smallest form measured about 0.5 cm, the largest specimen about 75 cm.

Ecology

The trilobites were predominantly epibenthic inhabitants of well-aerated, shallow coastal waters. Eyeless (blind) forms probably dug around in the mud as endobenthos. In contrast to other arthropods the trilobites had no differentiated mouth appendages, so that they will only have been able to eat minute organisms. Some probably fed on mud. Only a few phacopids are said to have had differentiated mouth parts which are taken to indicate a hunting lifestyle.

Trilobite provinces are distributed worldwide in the Lower and Middle Cambrian (Fig. 148).

Fig. 148. Marine faunal provinces in the Lower Cambrian based on trilobite finds: 1, *Nevadia*; 2, *Redlichia*; 3, *Holmia*; 4, *Fallotaspis*.

1.   Order: Agnostida

Very small forms, usually without eyes; cephalon and pygidium very similar; only two or three thoracic segments.

Examples:

Agnostus (Fig. 149, a): Two thoracic segments. Occurrence: Upper Cambrian.

Fig. 149. Order Agnostida: (a) *Agnostus*; (b) *Eodiscus*; (c) *Pagetia* (all × 0.33).

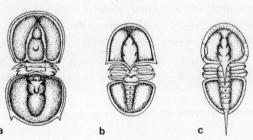

a   b   c

Eodiscus (Fig. 149, b): Three thoracic segments. Occurrence: Middle Cambrian.

Pagetia (Fig. 149, c): Two thoracic segments, proparian facial suture, eyes present, occipital and pygidial spines. Occurrence: Middle Cambrian.

2.   *Order: Redlichiida*
        Relatively large trilobites with opisthoparian or ankylosed (fused) facial suture. The semicircular cephalon often has genal spines. The glabella is clearly segmented. The eyes are usually large and more or less semicircular. The pygidium is very small. Some genera have a narrow opisthothorax.
        Examples:

Olenellus (Fig. 150, 1): Metaparian facial suture; 18–44 thoracic segments, opisthothorax. Characteristic of the 'Atlantic Province' (Europe, eastern part of North America, North Africa). Occurrence: Lower Cambrian (cf. p. 190).

Fig. 150. Order Redlichiida: (1) *Olenellus* (a: × 0.43; b: × 1); (2) *Holmia* (× 0.5).

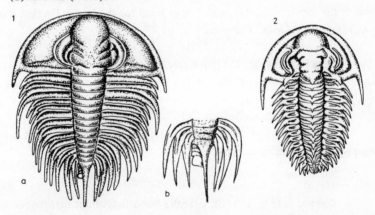

Holmia (Fig. 150, 2): Similar to *Olenellus*, but opisthothorax absent; genal spines and small axial spines; surface finely reticulate. Atlantic Province. Occurrence: Lower Cambrian.

Ellipsocephalus (Fig. 151, a): Subcylindrical; glabella triangular at the front, rounded cephalon; 12–14 thoracic segments; very small pygidium. Atlantic Province. Occurrence: Upper Lower Cambrian.

Fig. 151. Order Redlichiida: (a) *Ellipsocephalus* (×1); (b) *Paradoxides* (×0.25, with hypostome in upper figure, ×1.5); (c) *Redlichia* (×1); (d) *Ogygopsis* (×0.7).

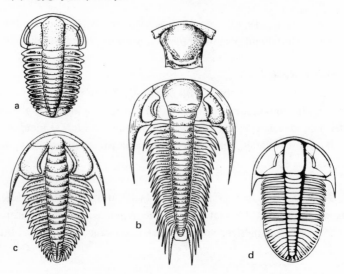

Redlichia (Fig. 151, c): Opisthoparian facial suture; glabella narrowing towards the front; no ocular ridge. Guide fossil for the Lower Cambrian of the 'Pacific Province' (Asia, Australia).

Paradoxides (Fig. 151, b): Glabella broad in front, long genal spines; no ocular ridge; very small pygidium. Atlantic Province. Occurrence: Middle Cambrian.

3.   Order: Corynexochida

The members of this small order are similar to the olenellids, but have larger pygidia and often dorsal spines.

Example:

Ogygopsis (Fig. 151, d): Glabella smooth, without furrows; ocular ridges present; eight thoracic segments; large pygidium composed of many segments. Occurrence: Middle Cambrian of western North America.

4.   Order: Ptychopariida

This is the most diverse trilobite order. The ptychopariids usually have opisthoparian facial sutures; the glabella is indented by deep lateral furrows. The glabellar furrows are directed backwards. Occurrence: Early Cambrian–Middle Permian.

Examples:

Ptychoparia (Fig. 152, a): Glabella tapering towards the front and furrowed towards the rear; straight ocular ridges; the border of the cephalon is convex; deeply furrowed pleura. Occurrence: Middle Cambrian.

Euloma (Fig. 152, b): Very deep glabellar furrows sharply curved backwards; index fossil for the Lower Ordovician.

Conocoryphe (Fig. 152, c): Eyes absent; very narrow librigenae; axis clearly segmented to the end of the tail. Atlantic Province. Occurrence: Middle Cambrian.

Fig. 152. Order Ptychopariida: (a) *Ptychoparia* (×0.33); (b) *Euloma* (×1.5); (c) *Conocoryphe* (×0.5); (d) *Dikelocephalus* (cephalon ×0.6, pygidium ×0.4); (e) *Olenus* (×1); (f) *Asaphus* (×0.75).

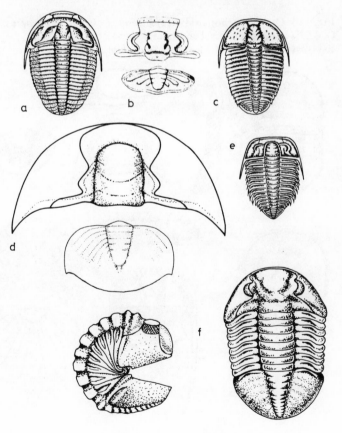

Dikelocephalus (Fig. 152, d): Similar to *Paradoxides*, but glabella not broad at the front and pygidium very large with two points; large pre-glabellar area. Guide fossil for the Upper Cambrian of the Pacific Province.

Olenus (Fig. 152, e): Small, fairly primitive; well defined straight ocular ridge; genal spines; small pygidium. *Olenus* is the best known genus of the Family Olenidae which includes most of the guide fossils for the Middle and Upper Cambrian of the Atlantic Province, e.g. *Parabolina*, *Eurycare*, *Peltina*.

Asaphus (Fig. 152, f): Large trilobite measuring up to 40 cm with wide doublure, ill-defined glabella, eight thoracic segments and large to very large pygidium. Diverse group. Occurrence: Ordovician.

Illaenus (Fig. 153, a): Similar to *Asaphus*, but glabella, axis and pygidium even less distinct; usually ten thoracic segments. Occurrence: Ordovician

Fig. 153. Order Ptychopariida: (a) *Illaenus* (× 0.75); (b) *Proetus* (× 0.5); (c) *Harpes* (dorsal and lateral view; × 0.55 and × 0.5, respectively); (d) *Trinucleus* (× 1.3); (e) *Scutellum* (× 0.4).

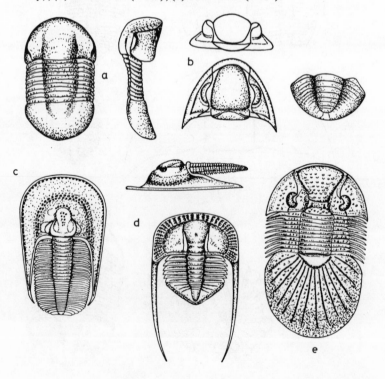

–Lower Silurian. *Bumastus* from the Upper Silurian and *Scutellum* (Fig. 153, e) from the Silurian/Devonian are similar.

Proetus (Fig. 153, b): Small; glabella relatively large, inflated and in some subgenera with granulation; very narrow fixigenae; cephalon with a pronounced marginal thickening. Occurrence: Ordovician-Devonian. Of all the trilobite families the Family Proetidae survived over the longest period of time: Upper Cambrian-Middle Permian.

Harpes (Fig. 153, c): Hypoparian facial suture; well defined ocular ridges; fringe with long genal spines; small thorax and pygidium. Occurrence: Silurian-Devonian.

Trinucleus (Fig. 153, d): Hypoparian facial suture; no eyes; very long pointed genal spines extending well beyond the pygidium. Occurrence: Ordovician.

5.   *Order: Phacopida*
Proparian or gonatoparian facial sutures, glabella broad at the front. The thorax consists of 8–19 segments. Occurrence: Lower Ordovician-Upper Devonian.
     Examples:

Phacops (Fig. 154, a–d): With tubercles; glabella inflated; glabellar furrow much reduced; schizochroal eyes. Tendency towards a reduction of the eyes and ankylosis (fusion) of the (proparian) facial suture. There is an additional ring between the glabella and the occipital ring. 'Salterian moulting' common. Occurrence: Silurian-Devonian.

Fig. 154. Order Phacopida: *Phacops*. (a), (b) Cephalon (× 0.7); (c) pygidium of an enrolled specimen, with cephalic doublure (× 1.3); (d) hypostome (× 0.6).

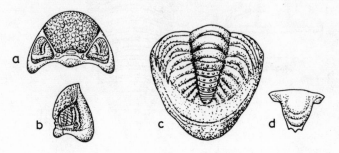

Fig. 155. Order Phacopida: (a) *Cheirurus* (× 0.6); (b) *Calymene* (× 0.4); (c) *Dalmanitina* (× 0.6).

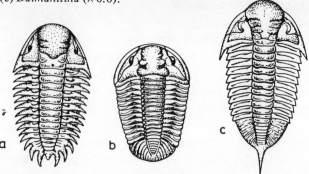

Dalmanitina (Fig. 155, c): Similar to *Phacops*, but without tubercles and with genal spines; pygidium sometimes with a spine at the end. Occurrence: Ordovician, ? Silurian. *D. socialis* (illustrated) is the guide fossil for the Middle Ordovician of Czechoslovakia.

Cheirurus (Fig. 155, a): Tuberculate; the rear glabellar furrows form an X with the neck furrow; pygidium with wide peripheral spines, similarly the pleura. Occurrence: Ordovician–Devonian.

Calymene (Fig. 155, b): Gonatoparian facial suture with tendency towards proparian suture; glabella with pronounced rounded and more or less distinct lobes. Occurrence: Ordovician–Devonian.

Homalonotus (Fig. 156, a–d): Transition from *Calymene*: cephalon and pygidium smoothed out, axis extremely wide. Characteristic trilobite of the

Fig. 156. Order Phacopida: *Homalonotus*. (a), (b) Dorsal and anterior view of the cephalon (× 0.3); (c) thorax with pygidium (× 0.35); (d) *hypostome* (× 0.6).

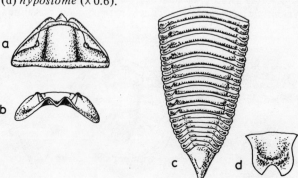

southern continents, but also known in the northern hemisphere. Occurrence: Silurian-Devonian.

6.   Order: Lichida
    Includes some extremely large trilobites such as the Genus *Uralichas* measuring 75 cm. Opisthoparian facial sutures. The glabella exhibits anomalous segmentation and is more or less strongly granulated. The cephalon is furnished with genal horns. Occurrence: Lower Ordovician-Upper Devonian.
    Examples:

    Lichas and *Ceratarges* (Fig. 157, a): Aberrant glabella. Occurrence: Ordovician-Silurian.

Fig. 157. Order Lichida: (a) *Ceratarges* (× 1.6). Order Odontopleurida: (b) *Odontopleura* (× 0.7).

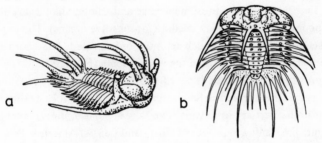

7.   Order: Odontopleurida
    Small trilobites with numerous spines on the cephalon, thorax and pygidium. Opisthoparian facial sutures. Occurrence: Middle Cambrian-Upper Devonian.
    Example:

    Odontopleura (Fig. 157, b): Widespread in the Upper Silurian of Europe.

2.   **Subphylum: Chelicerata**

    Chelicerates have one pair of jointed pincers (chelae) on the first pair of appendages (chelicerae) in front of the mouth which serve to grasp food. These are followed by a pair of pedipalps and four pairs of walking legs which may also terminate in chelae. Antennae are totally absent.

The chelicerates also differ from other arthropods by the possession of a prosoma (originally consisting of the cephalon and six additional segments). Behind this is the opisthosoma which may be divided into a broad anterior mesosoma and a narrower rear metasoma. This may be followed by a rudder-like or pointed telson.

The exoskeleton consists of a chitin-like material which is usually only sparsely strengthened with calcium salts, so the conditions of preservation are less favourable than in the case of the trilobites.

1.   Class: Merostomata
    The Merostomata include the Subclasses Xiphosura (relatives of the Recent Genus *Limulus*) and Eurypterida which boast the largest known arthropods.

1.   Subclass: Xiphosura (horseshoe crabs)
1.   Order: Xiphosurida
    In the xiphosures the prosoma is large and trilobite-like; there are lateral compound eyes and median ocelli (light-sensitive organs). The opisthosoma consists of nine segments which are more or less fused. The sword-shaped telson (terminal spine) is longer than the opisthosoma (Fig. 158).

Mesolimulus from the Jurassic is a well known fossil genus. Its type species *M. walchi* was described from the Solnhofen limestone (Malm ʓ). Quite similar genera occur in the Tertiary of Europe. The Recent genus *Limulus*, a 'living fossil', inhabits the shallow sea areas of North and Central America. Two similar genera (*Carcinoscorpius* and *Tachypleus*) occur in east and south-east Asia. Occurrence of the order: Silurian–Recent.

Fig. 158. Subclass Xiphosura: *Limulus,* Recent. Left, ventral view; right, dorsal view (× 0.5).

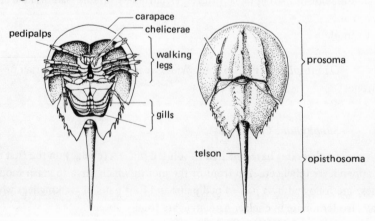

2.   *Order: Aglaspida*

The Aglaspida are small 2–6 cm long xiphosures with 11–12 opisthosomal segments which are not fused. They are restricted to the Cambrian and Ordovician. *Aglaspis* is the main genus.

According to recent studies, the Aglaspida are no longer to be assigned to the Chelicerata; their taxonomic status is quite uncertain.

2.   Subclass: Eurypterida

The eurypterids have a narrow exoskeleton with a relatively unsegmented prosoma. The opisthosoma consists of 12 articulated segments of which seven are assigned to the mesosoma and five to the metasoma. The eurypterids grew up to 2 m in length, which makes them the largest known arthropods. The last pair of prosomal appendages are modified into large paddles for swimming.

The presence of strong teeth on the proximal portion of the appendages enabled them to overpower larger prey. The eurypterids may well have been serious enemies of the contemporaneous vertebrates. From the Ordovician to Permian they first lived in the sea, but then invaded brackish water and fresh water and were even able to live on land for short periods.

Example:

Eurypterus (Fig. 159): Specimens of *Eurypterus* were found in excellent preservation by Holm in 1898 in the Silurian dolomites of the island of Ösel, so that this genus has been described in fine detail. Male and female genital appendages were observed on the second abdominal segment.

Fig. 159. Subclass Eurypterida: *Eurypterus*. Left, dorsal view; right, ventral view (× 0.3).

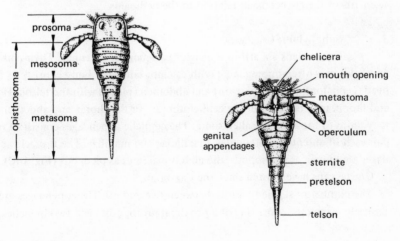

2.   Class: Arachnida (spiders and scorpions)
          As in the case of the Merostomata, the arachnid body is divided into
prosoma and opisthosoma, but the division is usually even more distinct. The six
pairs of appendages are also the same: one pair of chelicerae, one pair of
pedipalps and four pairs of walking legs. The Arachnida have no antennae or
mandibles.
          The main difference between the arachnids and the other chelicerates is
their terrestrial (occasionally secondarily aquatic) lifestyle. Gills are thus absent
and replaced by lung-books or tracheae. The arachnids also discard their chitinous
integument during moulting (ecdysis), as indicated by finds of exuviae in
Tertiary amber and in the Carboniferous. The chelicerae are often poisonous,
and silk glands are common. Only fluid food can be ingested; predigestion takes
place outside the body.
          The arachnids differ from insects by the possession of eight walking legs and
the absence of wings and a segmented head.
          Arachnids moult five to eight times during their phase of growth. They
measure between 0.5 and 200 mm. Most have a solitary carnivorous lifestyle.
          Because of their terrestrial mode of life, arachnids rarely occur as fossils,
and their phylogeny is thus badly documented. They are well represented only
in the Carboniferous and in Tertiary amber. It is evident that morphological
differentiation was decisively concluded as early as the Palaeozoic. The separa-
tion of the scorpions began in the Silurian and was completed in the Carboniferous,
after which the most specialised forms died out. The spider body plan was also
more or less developed in the Carboniferous. The oldest order (Scorpionida)
was already represented in the Silurian.
          The basic classification of the Arachnida is founded on the type of connec-
tion between the prosoma and the opisthosoma and on the position of the basal
segments of the chelicerae in relation to the pedipalps.

3.   Subphylum: Crustacea
          Crustaceans are arthropods with two pairs of antennae. Their chitinous
exoskeleton is often impregnated with calcium salts. The body consists of the
head (cephalon), thorax (pereion) and abdomen (pleon) with the telson. Head
and thorax may be fused into a cephalothorax, or the thorax and abdomen
may unite to form the so-called soma. The cephalic appendages consist of one
pair each of antennae, antennules, maxillulae and maxillae. The thoracic appendages are known as pereiopods, the abdominal ones as pleopods (Fig. 140).
          Crustaceans have existed since the Cambrian.
          The number of segments varies between five and 60. The appendages are
basically biramous, consisting of a proximal protopodite and two branches, the

endopodite and the exopodite. Two lateral branches on the protopodite may be modified as gills: the epipodite and pre-epipodite.

The carapace is a cuticular structure formed by a dorsal fold of the integument which arises from the posterior border of the cephalon. It is chitinous and largely impregnated with calcium salts. It covers the cephalon, and may extend over the thorax and completely fuse with it. The ostracod carapace envelops the body as a two-valved shell.

Crustaceans inhabit marine, brackish and fresh-water habitats and some are even secondarily terrestrial. Fossil forms are known since the Cambrian.

Classification of the Crustacea

Class: Cephalocarida (Recent)
Class: Branchiopoda (Lower Devonian–Recent)
Class: Mystacocarida (Recent)
Class: Euthycarcinoida (Triassic)
Class: Copepoda (Miocene–Recent)
Class: Ostracoda (Lower Cambrian–Recent)
Class: Branchiura (Recent)
Class: Cirripedia (Upper Silurian–Recent)
Class: Malacostraca (Lower Cambrian–Recent)

Of the nine classes tabulated here, the ostracods, cirripedes and malacostracans also have palaeontological significance; they will thus be dealt with briefly.

1.   Class: Ostracoda

Ostracods are small crustaceans, usually measuring between 0.5 and 5 mm, although some Palaeozoic ostracods grew to 30 mm in length. The body and its appendages are completely enclosed by a bivalved shell, the carapace, which is commonly extensively calcified. The two valves, which are of unequal size in some forms, are held together on the dorsal side by a ligament and often also by a hinge. The body is indistinctly segmented. The abdomen is rudimentary and furnished with a furca (anal bristles). Respiration occurs over the entire body surface.

The carapace consists of a left and right valve. There are one or more adductor muscle scars in the front part of the valves (Fig. 160*A*).

Along the dorsal margin, the hinge line, there are either numerous uniform teeth (taxodont hinge) or two teeth on one valve and two corresponding pits on the other valve (merodont hinge). If both valves have teeth, the hinge is known as an amphidont hinge. Dysodont hinge types have a hinge groove in each valve (Fig. 160*C*).

Fig. 160*A*. Class Ostracoda: schematic representation of a palaeocopid ostracod (left) and organisation of a living ostracod (right).

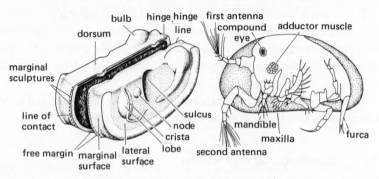

The carapace consists of three layers (Fig. 160*B*): (*a*) a thin chitinous external layer (rarely preserved in fossils); (*b*) an outer calcitic layer which is penetrated by pore canals; (*c*) an inner layer which may be calcified in the front ventral and the rear portions.

The external sculpture of the valves is highly variable, with a wealth of pits, tubercles, humps, ridges and spines. The sculpture can even vary within one species (different stages of moulting).

Many ostracods exhibit marked sexual dimorphism which is often recognisable by the presence of brood pouches in the females.

Fig. 160*B*. Pore canals and wall structure of the peripheral part of a podocopid ostracod valve.

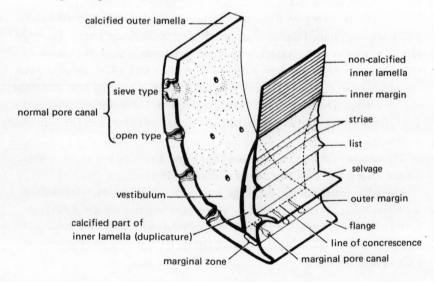

Fig. 160C. Some ostracod hinge types (dorsal view, right and left
valves): (1) adont, (2) prionodont, (3) lophodont, (4) palaeomerodont,
(5) holomerodont, (6) antimerodont, (7) hemimerodont, (8) entomo-
dont, (9) lobodont, (10) paramphidont, (11) hemiamphidont, (12) holo-
amphidont, (13) schizodont, (14) gongylodont.

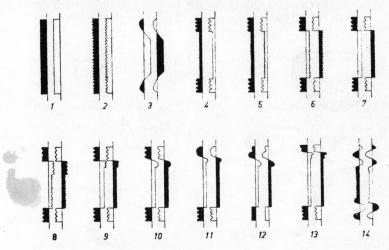

The majority of ostracods live in the sea where they are part of the benthos,
but they also inhabit brackish water and fresh water. Most feed on diatoms,
small animals and detritus. After the copepods, the ostracods are the most
abundant crustaceans in the present-day seas.

Their wide distribution, rapid evolution and relatively easy identification
have made the ostracods valuable guide fossils. After the foraminifera, the
ostracods are the next most important microfossils.

1.   *Order: Archaeocopida*
        Carapace thin, flexible, porous, made up of chitinous or phosphatic
material. Straight dorsal margin lacking hinge elements; ventral margin convex.
No muscle scars. The Archaeocopida probably lived in shallow waters. A pelagic
existence is also suggested by their occurrence in Cambrian pelagic limestones.
Occurrence: Lower Cambrian–Ordovician.

        Bradoria (Fig. 161, a): Carapace subquadrate, smooth or wrinkled;
prominent anterodorsal eye tubercles.

2.   *Order: Leperditicopida*
        Carapace thick, calcified; large and distinctive muscle scars. Dorsal
margin long and straight, often ending in cardinal angles. Adont hinge. Size:

Fig. 161. Class Ostracoda: (a) *Bradoria* (× 8); (b) *Beyrichia*, right valve
(× 17.5), adont hinge; (c) *Kloedenella*, right valve (× 20), transverse
section (× 45); (d) *Healdia*, right valve (× 67), dorsal view (× 67);
(e) *Cytherella*, left and right valve (× 27), dorsal view (× 27); (f) *Bairdia*,
right valve (× 43), dorsal view (× 43); (g) *Cypris*, interior (× 16), exterior
(× 13); (h) *Cypridea*, interior of left valve (× 30), exterior and dorsal
view (× 20); (i) *Cypridina*, exterior of left valve (× 20), interior (× 33).
LV, left valve, RV, right valve.

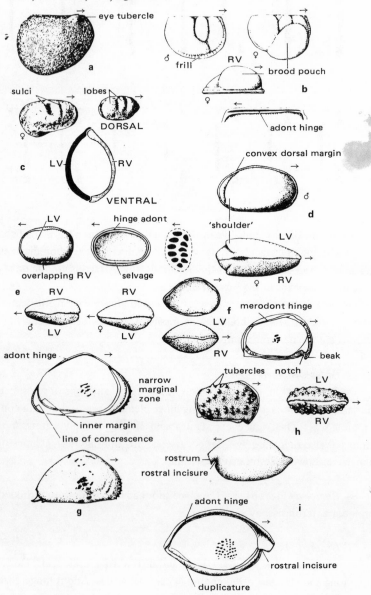

5–30 mm long. Eye spots and smooth valves with convex ventral margins indicate a shallow-water, pelagic existence. Occurrence: Ordovician–Devonian.

Leperditia: Carapace purse-shaped, smooth or punctate; distinct eye tubercles and adductor muscle scars. Occurrence: Lower Silurian–Upper Devonian.

3.   Order: Palaeocopida
.   Long straight hinge line, lobate and sulcate sculpture. Often distinctive sexual dimorphism expressed by the development of ventral brood pouches (crumina) in the female shells (so-called heteromorphs, as opposed to the shells of juvenile and male specimens which are collectively known as teknomorphs). Occurrence: Lower Ordovician–Middle Permian.

1.   Suborder: Beyrichicopina
Beyrichia (Fig. 161, b): Three distinct lobes and a granular or pitted surface; adont hinge. The heteromorph has globular brood pouches. Occurrence: Lower Silurian–Middle Devonian.

2.   Suborder: Kloedenellocopina
Kloedenella (Fig. 161, c): Subovate valves of unequal size, with the left valve overlapping the right one. Both valves bear two prominent anterodorsal sulci. Occurrence: Silurian–Devonian.

4.   Order: Podocopida
Dorsal margin convex or straight, ventral margin often concave but may be straight or convex. Occurrence: Lower Ordovician–Recent.

1.   Suborder: Metacopina
Occurrence: Ordovician–Lower Cretaceous.

Healdia (Fig. 161, d): Carapace smooth, rounded with 'shoulders' directed backwards near the posterior end. Occurrence: Devonian–Permian.

2.   Suborder: Platycopina
Occurrence: Triassic–Recent.

Cytherella (Fig. 161, e): Carapace smooth and ovate. Occurrence: Jurassic–Recent.

3.   Suborder: Podocopina
        Occurrence: Lower Ordovician–Recent.

Bairdia (Fig. 161, f): The large left valve partly overlaps the margins of
the right one; the hinge is adont. Occurrence: Ordovician–Recent.

Cypris (Fig. 161, g): Carapace smooth, subtriangular and relatively
large (up to 2.5 mm long); fresh-water ostracod. Occurrence: ?Jurassic,
Pleistocene–Recent.

Cypridea (Fig. 161, h): Dorsal and ventral margins relatively straight;
merodont hinge; surface pitted and pustulous; fresh-water to brackish-water
ostracod. Occurrence: Upper Jurassic–Lower Cretaceous.

5.   Order: Myodocopida
        Dorsal margins commonly convex. In the Superfamily Cypridinacea
there is a prominent anterior 'rostrum'. The ventral margin is convex. Occurrence: Ordovician–Recent.

Cypridina (Fig. 161, i): Prominent anterior rostrum and rostral incisure;
adont hinge. Occurrence: Upper Cretaceous–Recent.

2.   Class: Cirripedia (barnacles)
        As a result of their sessile lifestyle the cirripedes have produced aberrant
but nevertheless very successful forms. They are often fixed to foreign objects by
their head and enveloped by a skin-like mantle and by calcareous plates. These
days, they are distributed worldwide in all seas from the coastal areas down to
great depths. They have also developed a number of parasitic forms.
        Two orders have achieved greater palaeontological significance:

1.   Order: Acrothoracica
        In the Acrothoracica the females bore passages into limestones and the
calcareous shells of molluscs; they are usually accompanied by dwarf males.
Since they do not secrete calcareous shells themselves, the only fossil evidence
for their existence is their endolithic burrow. Their characteristic comma-shaped
bore holes are known from the Carboniferous to Recent.
        Example:

Rodgerella. Occurrence: Jurassic–Tertiary (Pliocene).

2.   Order: *Thoracica*
    The Thoracica live attached to the substrate, and their mantle is
strengthened by calcareous plates. This order includes the present-day barnacles,
which are very diverse and widespread. They are attached either by a broad base
or by a stalk; the plates may remain isolated or fuse into virtual shells.
    The Thoracica are known to have existed for certain since the Upper Silurian.
Examples:

    Balanus (Fig. 162, a, b): Shell consisting of six firmly connected
calcareous plates on a calcareous base. The plates are hollow and subdivided
by walls and parietal tubes. Occurrence: Oligocene–Recent.

Fig. 162. Class Cirripedia: *Balanus*. (a) Organisation (schematic),
(b) arrangement of the plates (schematic).

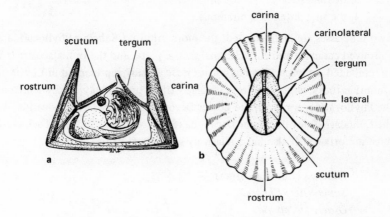

    Lepas (Fig. 163): Shell consisting of five plates. Attached to drifting
objects by a fleshy stalk; hemiplanktonic. Occurrence: Eocene–Recent.

3.   Class: Malacostraca
    The Malacostraca embrace the more highly organised crustaceans which
are closely related by having a number of features in common. In spite of the
considerable differences in size and shape, they share the following characteristics:

    1. the number of body segments is usually 21 (head with six segments,
        thorax with eight, abdomen with six segments and the telson);

Fig. 163. Class Cirripedia: *Lepas.* Left, a diagram of its organisation; right, the arrangement of the plates (× 1).

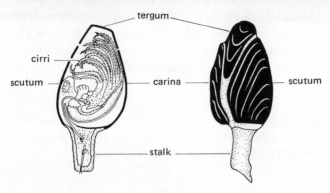

2. the carapace, a fold of skin emanating from the head, usually also envelops the thorax;

3. flexible stalked eyes are present;

4. the first antenna is biramous.

Two subclasses are distinguished: the more primitive **Subclass Phyllocarida** with a bivalved carapace and an additional rostral plate, and the Eumalacostraca with an undivided carapace. The Phyllocarida are already represented in Lower Cambrian rocks, while the first Eumalacostraca, presumably descended from the Phyllocarida, appear in the Lower Devonian with the Order Eocarida. Among the Eumalacostraca the Order Decapoda (Superorder Eucarida) has achieved the most importance. It also has a relatively good fossil record.

Subclass: Eumalacostraca

1.   Superorder: Eucarida

1.   Order: Decapoda

The first three pairs of thoracic appendages (thoracopods) of the decapods are specialised as maxillipeds for the uptake of food, so that only the rear five pairs of appendages serve the purpose of locomotion as pereiopods or pleopods (for swimming): hence the name decapods. The first of these five pairs is usually particularly well developed and furnished with pincers (Fig. 140).

The carapace is fused with the thoracic segments and divided into areas by the presence of furrows. The arrangement of these furrows is of considerable taxonomic importance for fossil decapods, since carapace finds are especially common.

In primitive (swimming) decapods the abdomen is developed normally, with pleopods being used for locomotion. It is gradually reduced in the course of phylogeny and is finally curved under the thorax in short-tailed decapods.

Views on decapod classification have changed many times over. In a first classification, the mode of life was used as a criterion, and the following groups were accordingly distinguished: Natantia (swimming decapods) and Reptantia (crawling decapods). The former included mainly the Macrura (long-tailed decapods), the latter the Brachyura (short-tailed decapods) to a greater or lesser extent.

In addition, the gill type was also taken into consideration: trichobranchs, filament-shaped gills; phyllobranchs, leaf-shaped gills; dendrobranchs, dendritic gills. The number of pincer-bearing pairs of legs served to distinguish the Trichelida (with pincers on three pairs of legs) from the Heterochelida (without pincers on the third pair of legs). More recent classifications are attributable to Beurlen and Glaessner in particular. The classification used here and in the *Treatise on Invertebrate Paleontology* (Moore & Teichert, 1952ff) originates from Glaessner.

1. Suborder: Dendrobranchiata

Swimming forms, dendrobranchiate, pincers on the first three pairs of pereiopods. Occurrence: Permo-Triassic–Recent.

Example:

Aeger (Fig. 164): Long-tailed decapod with long laterally granulated rostrum; pereiopods with one to three spines. Occurrence: Upper Triassic–Upper Jurassic.

2. Suborder: Pleocyemata

Decapods with gills which are not secondarily branched. Occurrence: Permo-Triassic–Recent.

Fig. 164. Class Malacostraca (Order Decapoda): *Aeger* (× 0.4).

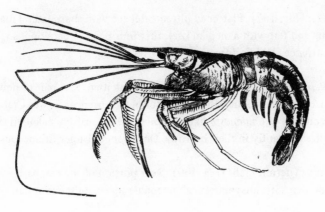

Fig. 165. Class Malacostraca (Order Decapoda): *Mecochirus* (× 0.5).

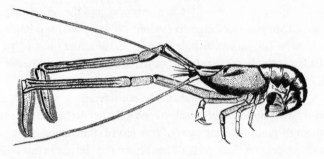

Mecochirus (Fig. 165): Thin carapace with short anterior portion; first pair of pereiopods extremely elongate and slender. Occurrence: Middle Triassic–Upper Cretaceous.

Pemphix (Fig. 166): Cylindrical decapod with pronounced sculpturing; rostra long and spoon-shaped; first pair of pereiopods well developed, second to

Fig. 166. Class Malacostraca (Order Decapoda): *Pemphix* (× 0.6).

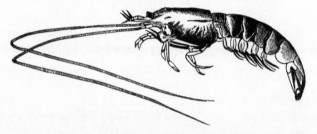

fifth pairs more or less uniformly developed; well developed abdomen with transverse furrows and pointed pleura. Occurrence: Middle Triassic.

Eryon (Fig. 167): Flattened pitted carapace with sharp lateral margins; abdomen long and flat with a median keel; first four pairs of pereiopods with pincers. Occurrence: Middle Jurassic–Lower Cretaceous.

Callianassa (Fig. 168): Carapace weakly calcified; first pair of pereiopods unequal in size (heterochelous). Burrowing form. Occurrence: Lower Cretaceous–Recent. The pincers of *Callianassa*-like decapods are extensively calcified and thus fossilised quite frequently in their burrows. Occurrence: Upper Cretaceous–Recent.

Pagurus (hermit crab) (Fig. 169): Rear portion of the carapace weakly calcified; abdomen soft, unsymmetrical; unequal pincers on the first pair of

Fig. 167. Class Malacostraca (Order Decapoda): *Eryon* (× 0.33).

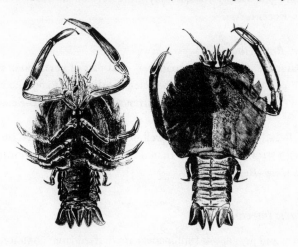

Fig. 168. Class Malacostraca (Order Decapoda): *Callianassa* (× 1).

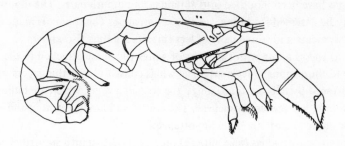

Fig. 169. Class Malacostraca (Order Decapoda): *Pagurus* (× 2).

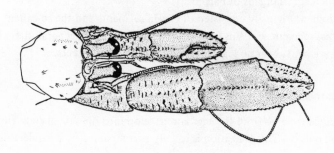

pereiopods, the right one usually considerably more developed. Occurrence:
Lower Cretaceous-Recent.

3.   Suborder: Brachyura

Crabs. Carapace increasingly broadened; first pair of pereiopods always
with pincers, third pair always without. Abdomen short and flat and tucked
under the sternum (ventral integument). Occurrence: Lower Jurassic-Recent.

Dromia: Relatively small and not well differentiated; rounded carapace;
first pereiopods with strong pincers, fourth and fifth pairs small and upturned.
Occurrence: Palaeocene-Recent.

4.   Subphylum: Insecta

The insects and myriapods (millipedes) are derived from common
ancestral forms which probably closely resembled the myriapod class of the
Symphyla. The two taxa are collectively known as the Mandibulata.

The bodies of insects are divided into three distinct regions: the head (caput),
the thorax and the abdomen. The head consists of seven fused segments whose
appendages have been modified into antennae and mouth parts. The mouth
parts may be designed for biting, sucking or impaling. The thorax is made up
of three segments with three pairs of legs (insects are thus also known as
Hexapoda) and as a rule with two pairs of wings which are attached to the
second and third thoracic segments. The wings consist of two thin layers of
chitin stiffened by net-like thickenings, the so-called veins, which are hollow
and contain blood vessels or tracheae. The course and arrangement of the
main veins have great taxonomic importance.

The front pair of wings (alae anticae) may be modified into strengthened
wing covers (elytra) which cover the rear pair of wings in the resting position;
elytra are characteristic of the Coleoptera (beetles). In the Diptera (two-winged
insects), on the other hand, the rear wings are reduced to small club-shaped
halteres which act as organs of balance.

The abdomen originally consisted of eleven segments and the so-called
telson or anal segment, but this number may be reduced. In general, the
abdominal segments bear no leg-like appendages in adult insects.

Ontogeny

Larvae emerge from the eggs and develop into the sexual stage (imago)
after several moults. This transition is usually accompanied by considerable

morphological changes and the process is known as metamorphosis (metabolism). Development without metamorphosis (in wingless insects) is known as ametabolism. Continuous metamorphosis, in which the wings, for example, are formed in several stages, is called hemimetabolism (e.g. dragonflies and Hemiptera). In holometabolous insects, which have a complete metamorphosis, the imago stage is preceded by a pupal stage during which neither locomotion nor feeding are possible. Holometabolism is characteristic of butterflies and beetles, for example.

Fossil insects

While the number of modern insect species is estimated to be close to one million, there are only about 12 000 fossil species of which more than half are from the Tertiary Period. Their significance for the recognition of phylogenetic relationships is correspondingly small. Their small size and terrestrial lifestyle made their preservation as fossils difficult. The Upper Carboniferous and Permian as well as the Tertiary (lignites and amber especially) are therefore the best sources of insect fossils. A detailed study of fossil insects presupposes a knowledge of modern forms, and since a full discussion of the latter does not come within the scope of this book and since the fossil forms are of no great stratigraphic importance, only a few genera of general interest are covered.

Rhyniella from the Middle Devonian Rhynie cherts (Scotland) is regarded as the oldest and also the most primitive insect fossil. It belongs to the wingless Collembola (springtails).

Insects which are only able to fold their wings vertically with respect to the body are collectively known as the Palaeoptera. They are primitive in that they are not able to rotate their wings backwards across the abdomen. This group includes the present-day mayflies (Ephemerida) and dragonflies (Odonata), and the fossil Palaeodictyoptera and Megasecoptera, both of which could grow to huge sizes. The two latter groups are restricted to the Upper Carboniferous and the Permian.

The Palaeodictyoptera were large, primitive and probably quite heavy insects with a small head and non-folding wings attached by a broad base. Their mouth parts were designed for biting or for sucking plant juices. The largest known representative of the Palaeodictyoptera is *Stenodictya* (Fig. 170) from the Upper Carboniferous, with a wing-span of about 50 cm. It is characterised by the development of a pair of wing-like projections on the first thoracic segment and pleuron-like lateral projections on the abdominal segments.

The Megasecoptera, also known as Protodonata, are regarded as the ancestors of the Odonata (dragonflies) which first appear in the Triassic. They were slender carnivorous insects, resembling the present-day dragonflies, with a mobile head

Fig. 170. Subphylum Insecta: *Stenodictya* (× 0.4).

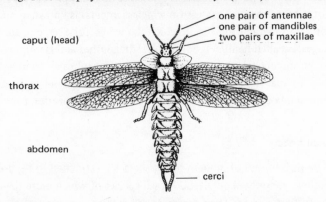

Fig. 171. Subphylum Insecta: *Meganeura monyi* (× 0.15).

and large eyes and the capacity to fly very fast. The thoracic and abdominal segments were quite similar. Their rigid wings were attached by small bases. The largest of these primaeval dragonflies was *Meganeura monyi* from the Upper Carboniferous (Upper Stephanian), with a maximum wing-span of 70 cm (Fig. 171).

Phyletic Group: Tentaculata, Lophophora

Sessile protostomes in which the mouth is surrounded by a ring of tentacles (lophophore). Usually U-shaped intestine; one pair of nephridia; oesophageal ganglion. The group includes the Phyla Phoronida, Bryozoa and Brachiopoda (Fig. 172), and here the Conodontophorida are also included.

Fig. 172. Phyletic Group Tentaculata: geological distribution and relative abundance.

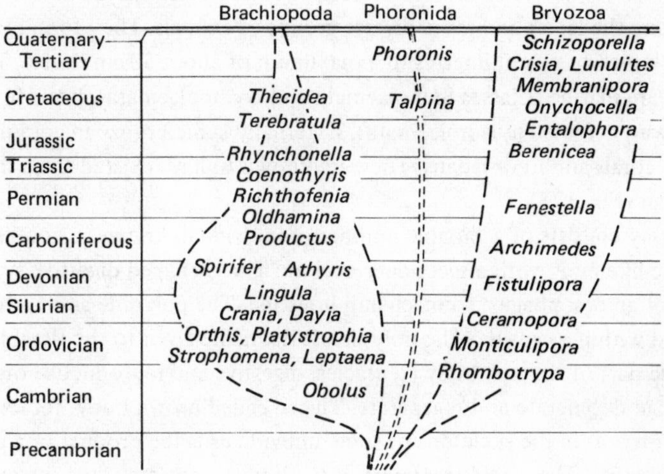

	Brachiopoda	Phoronida	Bryozoa
Quaternary		Phoronis	Schizoporella
Tertiary			Crisia, Lunulites
Cretaceous	Thecidea	Talpina	Membranipora
	Terebratula		Onychocella
Jurassic			Entalophora
Triassic	Rhynchonella		Berenicea
	Coenothyris		
Permian	Richthofenia		
	Oldhamina		Fenestella
Carboniferous	Productus		Archimedes
Devonian	Spirifer Athyris		Fistulipora
Silurian	Lingula		
	Crania, Dayia		Ceramopora
Ordovician	Orthis, Platystrophia		Monticulipora
	Strophomena, Leptaena		Rhombotrypa
Cambrian	Obolus		
Precambrian			

13. Phylum: Phoronida

The Phoronida have few species. They are characterised by a tube-like body measuring only a few centimetres and a horseshoe-shaped lophophore. The epidermis secretes a chitinous envelope which is usually covered with foreign particles.

The two living phoronid genera *Phoronis* and *Phoronopsis* are distributed worldwide and live in dense colonies on muddy or sandy substrates or else they bore into rocks and mollusc shells.

Since the Phoronida very probably represent an old residual group, their fossil record is of great interest. If we assume that the extinct phoronids inhabited the same biotopes as their present-day representatives, their tubes, bore-holes or tunnels should be preserved in shells or rocks as ichnofossils. As there seem to be no unambiguous morphological criteria for fossil phoronid burrows, evidence for this phylum is hard to find. The interpretation of the *Scolithos* tubes in the Lower Palaeozoic sandstones and of tube-shaped ducts in the Mesozoic–Cenozoic limestones as phoronid burrows was already considered decades ago. Only recently, E. Voigt was able to demonstrate the conformity between the fossil ichnogenus *Talpina*, which is particularly well known from belemnite rostra of the Cretaceous chalk and from Tertiary mollusc shells, and the burrows of Recent *Phoronis* species.

The Phylum Phoronida is thus certain to have existed since the Mesozoic, but it is very likely that it also existed in the Devonian.

14. Phylum: Bryozoa (Polyzoa; Ectoprocta)

With about 4000 Recent and about 16 000 fossil species, the Bryozoa rank among the larger and more diverse groups of animals. They form mostly sessile colonies a few millimetres to a maximum of about 50 cm in size. The majority are marine (Classes Stenolaemata and Gymnolaemata), but a few live in fresh water (Class Phylactolaemata). Externally some bryozoan colonies resemble corals and hydrozoans. They are known to have existed since the Ordovician (Fig. 173).

A colony consists of a variable number of individuals known as zooids. These have a sac-like body with a secondary coelom. The U-shaped digestive system consists of an oesophagus, stomach and intestine. The polypide and cystid are contained within a capsule. The polypide is the name given to the flexible and extensible part of the soft body (tentacles, digestive and reproductive organs, etc.). It can degenerate and regenerate. The so-called brown body, for example, which is present in the skeletons of some individuals is the product of a degenerated polypide. The cystid (rear part of the body) is the envelope surrounding the polypide, and it consists of organic material. The polypide can be repeatedly formed by the cystid. The mouth opening of the zooid is surrounded by a ring of tentacles which is joined at the base into a circular or horseshoe-shaped lophophore. The anus is outside the ring of tentacles near the mouth (Ectoprocta).

The tentacles, mouth and anus can be withdrawn into or extended out of the capsule by muscular action or by a hydrostatic apparatus. A vascular system is absent. The mostly calcareous skeleton is secreted by the epidermis. Bryozoans feed on microplankton by creating currents with their tentacles.

Fig. 173. Geological distribution of the bryozoan orders.

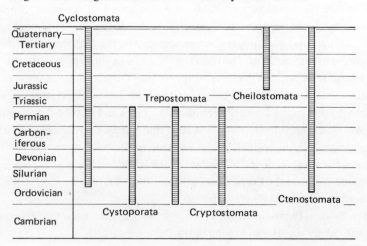

With very few exceptions, the bryozoans are hermaphrodites. Fertilisation usually occurs in the coelom. The eggs tend to mature in special brood chambers, the so-called ovicells. They develop into cyphonaute larvae which settle down on a solid substrate after a short free-swimming developmental phase, usually lasting only a few hours, and form the first individual of a colony, the ancestrula. Bryozoan larvae tend to exhibit a negative phototactic response, since they predominantly colonise the shaded undersides of substrates.

A bryozoan colony is known as a zoarium (plural: zoaria), the capsule of an individual as a zooecium (plural: zooecia). The zooecia arise by the process of budding. The opening of the zooecium is called the aperture or orifice; in forms of the Order Cheilostomata, for example, it may be closed by an operculum. The shape of the colonies is highly variable. In addition to the variety of tree-like colonies, there are encrusting colonies and unattached disc-shaped or plate-like colonies.

Sometimes there is a distinct polymorphism among the zooecia. A calcareous skeleton with a normal feeding individual is known as an autozooecium. Ceno-zooecia are smaller polymorphic calcareous skeletons without a polypide. Depending on their shape and position within the colonies of the Stenolaemata they are described as mesopores, acanthopores, firmatopores, nematopores, etc. In many cheilostome species there are so-called heterozooecia, in addition to the autozooecia, which are modified into avicularia and vibracularia (Fig. 174). The avicularia (singular: avicularium) can be interpreted as defensive zooecia. They are two-armed individuals without a polypide, resembling the shape of a bird's head, and with an operculum modified into a snapping organ (mandible). They are not preserved as fossils. In addition to the avicularia, the Cheilostomata have long bristles (setae), the so-called vibracula (singular: vibraculum), which

Fig. 174. Organisation of a bryozoan of the Order Cheilostomata.

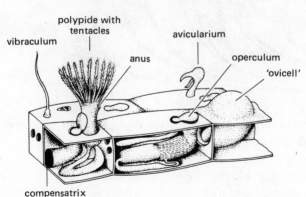

polypide with
tentacles

vibraculum

anus

avicularium

operculum

'ovicell'

compensatrix

are not calcified and which are only recognisable as pore-like places of attachment in fossils.

The classification of the Bryozoa is still partly in a rather unsatisfactory state, since important internal and external features such as the shape of the zooids, various pore canals, the ovicells and the architecture of the wall, are evaluated very differently in taxonomic terms by different authors.

1.   Class Stenolaemata

Marine bryozoans with calcified cylindrical zooecia which may be subdivided by diaphragms in some genera. Circular lophophore. Autozooecia and intercalated heterozooecia. Polymorphism. Ovicells very variable in shape. Occurrence: Ordovician–Recent.

1.   *Order: Cyclostomata*

Mostly tube-shaped, but occasionally polygonal zooecia. Various cenozooecia. The zooecia are in contact with one another by means of mural pores (Fig. 175). Very different, taxonomically important, ovicell types. Many tree-like and encrusting growth shapes are known. Occurrence: Ordovician–Recent.

Fig. 175. Organisation of a bryozoan of the Order Cyclostomata.

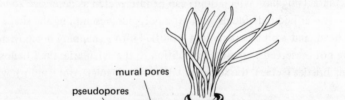

The Cyclostomata experienced their prime in the Upper Cretaceous; they have been on the decline ever since.

Stomatopora (Fig. 176, 4a): Colony of uniserial encrusting bifurcating branches. Zooecia arranged uniserially. The zooecia may stand upright distally

Fig. 176. Order Cyclostomata: (1) *Spiropora* (× 8); (2) *Entalophora* (× 8); (3) *Crisia* (× 10); (4a) *Stomatopora* (× 12); (4b) *Berenicea* (× 12).

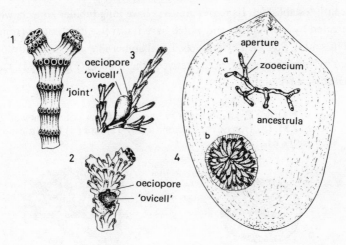

and form a peristome with a circular orifice. The ovicell is developed from an inflated bubble-shaped zooecium. Occurrence: Ordovician–Recent.

Berenicea (Fig. 176, 4b): Usually circular colonies 2–5 mm in diameter, encrusting in a single layer. Tube-shaped zooecia are arranged radially round the centre of the colony. Peristome with a circular cross-section. Since different types of ovicells may be developed, the colonies are assigned to different genera such as *Diaperoecia*, *Plagioecia* and *Oncousoecia*. All colonies without ovicells are assigned to the form-genus '*Berenicea*'. Occurrence: Ordovician–Recent.

Entalophora (Fig. 176, 2): Colony shaped like a tree-trunk, sometimes with branches; with narrow axial canal. Circular peristome which sometimes projects from the colony as a tube. Ovicell developed as a triangular, more or less heart-shaped brood chamber. Occurrence: Jurassic–Recent.

Crisia (Fig. 176, 3): Ramose colony; usually articulated. Zooecia tubular and in biserial arrangement; peristome only open to one side. Symmetrical sac-shaped ovicells which stand out clearly in parallel with the axes of the zooecia. Large terminal oeciopore. Occurrence: Tertiary–Recent.

Spiropora (Fig. 176, 1): Ramose colony. The orifices of the zooecia are arranged in the shape of a whorl. Various gonozooecia in the same colony; homeomorphism. Occurrence: Jurassic–? Tertiary.

2.   Order: Cystoporata

This order, which embraces about 50 genera from the Palaeozoic, was only recently established. Its representatives have long tubular zooecia which are connected by mural pores and which have transverse diaphragms (transverse calcareous septa). So-called cystopores, i.e. chamber-like supporting structures separated by transverse septa, occur between the zooecia (Fig. 177, d). Occurrence: Ordovician–Permian.

Fig. 177. Order Cystoporata: *Ceramopora.* (a) A colony viewed from above (× 3); (b) transverse section (× 20); (c) and (d) longitudinal sections (× 8).

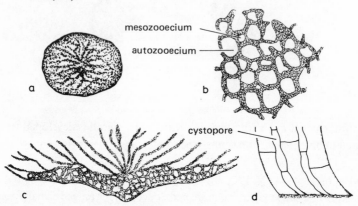

Ceramopora (Fig. 177, a–d). Colony disc- or bowl-shaped, encrusting, usually up to 1 cm in diameter. Tubular zooecia. On the surface of the colony the apertures with their hat-like lunaria are arranged in a radial pattern round a convex centre (macula). The mesopores are large and irregularly distributed. Occurrence: Ordovician–Devonian.

Fistulipora (Fig. 178, a, b): Colony thin lamellar to multilayered and massive; very variable in shape. The surface is covered with monticuli (hump-shaped mounds) or maculae (bowl-shaped depressions) or frequently with both. Zooecia more or less round. Convex, concave or straight diaphragms are visible in longitudinal sections. The interstitial tissue is narrow to wide-meshed. Lunaria of different shapes and sizes are commonly present with only one opening. Occurrence: Silurian–Permian.

3.   Order: Trepostomata

The Trepostomata occur in massive, partly reef-forming colonies of long tubular zooecia. There are no mural pores between the zooecia. The apertures

Fig. 178. Order Cystoporata: *Fistulipora*. (a) Transverse section; (b) longitudinal section (× 15).

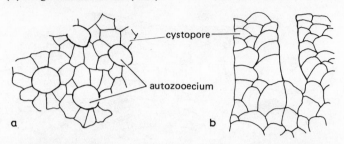

are polygonal. The zooecia have numerous diaphragms in the distal portion. About 115 genera have been described for the Palaeozoic, the majority of which occurred in the Ordovician. The zooecia often form hump-shaped mounds (monticuli) or bowl-shaped depressions (maculae) on the surface of the colonies. The very complicated internal architecture necessitates the use of thin sections for the purpose of identification (Fig. 179).

Fig. 179. Order Trepostomata: *Tabulipora*, three-dimensional representation of part of a colony.

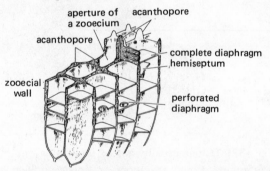

Monticulipora (Fig. 180): Colony massive and tuberous, sometimes forming blunt lateral branches. More or less regularly arranged monticuli on the surface. Thin-walled polygonal zooecia. In longitudinal sections convex cystiphragms and more or less straight diaphragms are visible. Apertures polygonal. Acanthopores small with granulated walls. Few mesopores with diaphragms. Occurrence: Ordovician.

Prasopora (Fig. 181): Colony discoidal to hemispherical; massive. In longitudinal section the zooecia exhibit overlapping cystiphragms which are slightly convex peripherally and connected by straight diaphragms. Small polygonal mesopores, tabulate. Occurrence: Ordovician.

Fig. 180. Order Trepostomata: *Monticulipora*. (a) A colony (×0.8); (b) peripheral longitudinal section (×25); (c) transverse section (×25).

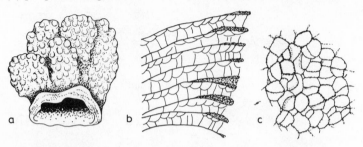

Fig. 181. Order Trepostomata: *Prasopora*. A colony viewed (a) from above, (b) from the side, (c) in longitudinal section; (d) the base of a colony (a–d: ×1); (e) transverse section (×12).

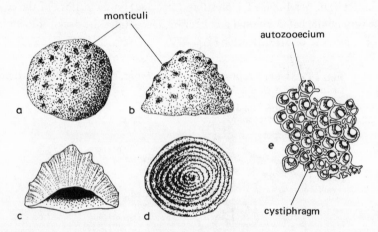

Dekayella (Fig. 182): Colony usually ramose; undulating. On the surface there are rows or groups of small acanthopores. Zooecia polygonal with straight diaphragms; cystiphragms absent. Many mesopores. Occurrence: Ordovician.

Rhombotrypa (Fig. 183): Colony free-standing from the base and branched. Zooecia consisting of prismatic tubes. In the axial region the zooecia are square in cross-section. Occurrence: Ordovician.

4. *Order: Cryptostomata*
     The colonies are mostly net-like or bush-shaped, but often also funnel-shaped. Acanthopores of variable size are developed on the surface of the colony. The zooecia are relatively short and tubular with diaphragms. The distal portion

Fig. 182. Order Trepostomata: *Dekayella.* (a) A fragment of a colony (× 0.8); (b) three-dimensional representation of part of the colony (× 3); (c) longitudinal section (× 20).

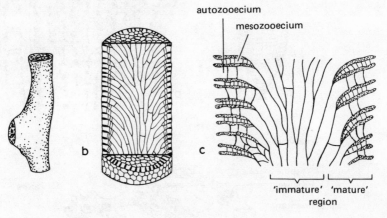

autozooecium

mesozooecium

'immature' 'mature'
region

Fig. 183. Order Trepostomata: *Rhombotrypa.* (a) Peripheral longitudinal section (× 5); (b) transverse section (× 3).

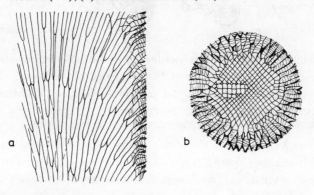

of the zooecia is funnel-shaped with a hemiseptum and extensively calcified. About 150 genera are known. Some cryptostomes are important as rock-formers. Occurrence: Ordovician–Permian.

Fenestella (Fig. 184): Colony funnel- or fan-shaped, net-like. Narrow branches which are arranged roughly in parallel are connected by cross-bars (dissepiments) constructed at more or less regular intervals. The resulting spaces are known as fenestrules (little windows). The branches consist of biserially arranged zooecia which terminate at the surface of the branch (inside the funnel-shaped colony) in circular apertures. The two rows of zooecia are separated by a median keel (carina) which may be furnished with thorn-like acanthopores.

Fig. 184. Order Cryptostomata: *Fenestella.* (a: × 2.5, b: × 12; c: × 25).

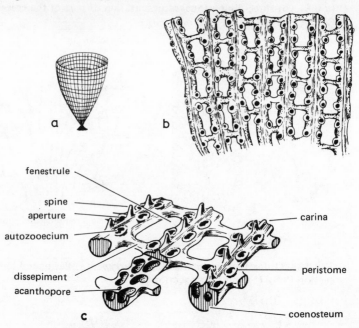

The fenestellids are distributed all over the world and are especially abundant in reef-like deposits of the Devonian and Permian. *Fenestella retiformis* among others participated in the formation of the reefs in the Germanic Zechstein (Upper Permian). Occurrence: Ordovician–Permian.

Archimedes (Fig. 185): The colony consists of several *Fenestella*-like funnels helically arranged round a central axis. The colonies may grow to heights of several tens of centimetres. Like *Fenestella*, the surface of the colony is

Fig. 185. Order Cryptostomata: *Archimedes* (a, b: × 1, c: × 2).

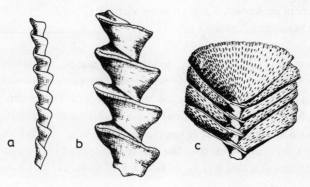

covered with two rows of apertures along the branches. Occurrence: Carboniferous–Permian.

Acanthocladia: Colony consisting of robust stem with short robust branches branching out obliquely. On the frontal side of the branches there are three or more rows of zooecia with round apertures; distal side usually with fine longitudinal striae. Occurrence: Carboniferous–Permian.

Rhombopora (Fig. 186): Slender, dichotomously branched dendritic colony. Peripheral area of the colony thick-walled (mature area). The zooecia exhibit only a few diaphragms in longitudinal section; no hemisepta. The oval '

Fig. 186. Order Cryptostomata: *Rhombopora* (a: ×1.5; b, c: ×10).

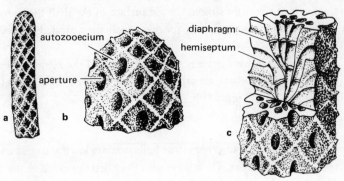

apertures terminate within an oblique hexagonal vestibulum; they are arranged in regular oblique rows. Each aperture is surrounded by very small acanthopores and possesses a medium-sized distal acanthopore. Mesopores are absent. Occurrence: Devonian–Permian.

2.   Class: Gymnolaemata
       Predominantly marine bryozoans with a circular lophophore. The zooecia are usually pitcher-shaped or box-like and nearly always calcified. There is polymorphism, which is sometimes quite pronounced. Class of fossil significance. About 650 genera are known from the Ordovician to Recent.

1.   Order: Ctenostomata
       Exclusively marine bryozoans with about 50 known genera. The zooecia are cylindrical or flattened and commonly connected by stolons; the walls are not calcified. The animals bore into calcareous substrates or encrust substrates or grow upwards. In addition to *Ropalonaria* from the Palaeozoic,

boring genera are represented by *Terebripora*, *Spathipora*, *Penetrantia* and *Immergentia*. Some encrusting ctenostome genera have been preserved by biomuration, i.e. the chitinous colonies were overgrown by encrusting bryozoans, serpulids, oysters, etc., and thus preserved as impressions in the fossil record. The relevant genera preserved in this way include *Arachnidium* and *Amathia*. Occurrence: Ordovician–Recent.

Boring, or rather mining ctenostome bryozoans are quite common on the smooth inner surface of mollusc shells, e.g. in oysters, while the ribbed outer surface is not attacked. The stolons always run in straight lines, usually just below the surface. Variation in the diameter of the stolon is frequently observed, giving it the appearance of a string of pearls. The sac-shaped or pear-shaped zooecia 'hang' from the stolon more or less symmetrically and are sunk into the substrate, or else they lie horizontally, more or less in the direction of the longitudinal axis of the main stolon. On the surface of the shell the round to oval or bowl-shaped cross-sections of zooecia which are corroded to varying degrees are visible.

The Recent genus *Bowerbankia* is an upright form of the ctenostomes. The oval, up to 1 mm long, zooecia are arranged semi-spirally round a central branch in groups of up to 20 and more (Fig. 187, d).

2. Order: Cheilostomata

With their 600 or so genera the Cheilostomata are the largest and most diverse bryozoan order. Their representatives first appeared at the end of the Jurassic. As a rule, cheilostome bryozoans have more or less box-like or pitcher-shaped calcified zooecia with a frontal aperture which may be closed by an operculum. There is pronounced polymorphism, with the presence of

Fig. 187. Order Ctenostomata: (d) *Bowerbankia*, expanded polyp (×5). Order Cheilostomata: (a)–(c) *Flustra*. (a) A colony (×0.5); (b) various zooecia (×10); (c) zooecia with an intercalated avicularium (×10).

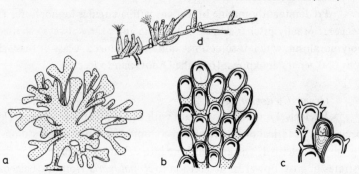

avicularia, vibracularia and a variety of ovicells (synonym: ooecia; singular: ooecium) (Fig. 174). The cheilostomes are the most highly developed and currently dominant bryozoans. In the coastal facies of the Upper Maastrichtian and in the Baltic Danian they are partly rock-forming. Occurrence: Jurassic–Recent.

Membranipora (Fig. 188, a): A typical collective genus with more than 600 species. A reclassification of this large form-group is still lacking.

Fig. 188. Order Cheilostomata: (a) *Membranipora* (×10); (b) *Schizoporella* (×6, ×15).

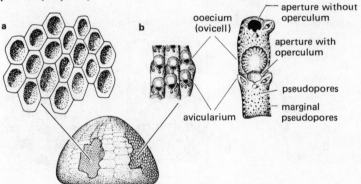

Colonies mostly encrusting or bilamellar, erect. Zooecia irregularly hexagonal to wide pear-shaped; 0.3–0.8 mm long and often as wide. The aperture (opesium) is medium-sized with the tendency towards a triangular shape, or large in which case it occupies the whole width of the zooecium. Frontal wall partly or wholly membranous. Avicularia variable in size. Outside on the distal end of the zooecium there is an ovicell (ooecium) which may extend across the distal portion like a helmet (hyperstomial type). Occurrence: Cretaceous–Recent.

Lunulites (Fig. 189): Colony unattached, discoidal or dome-shaped, up to a few millimetres in diameter. Zooecia arranged in radiating rows. Opesia

Fig. 189. Order Cheilostomata: *Lunulites* (×9).

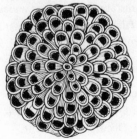

more or less semicircular; auriculate avicularia; vibracula variable in size. Ovicell developed within the zooecium (endozooecial). Occurrence: Cretaceous–Recent.

Schizoporella (Fig. 188, b): Colonies generally encrusting. Aperture (orifice) crescent-shaped with an operculum and a small proximal sinus (rimule) and a lateral avicularium. The zooecia exhibit largish pseudopores along the lower margin. Hyperstomial ovicells. Occurrence: Tertiary–Recent.

Coscinopleura (Fig. 190): Bifoliate colony of small compressed, dichotomously branched stems whose edges are formed by paired vibracularia touching

Fig. 190. Order Cheilostomata: *Coscinopleura* (a: ×2; b: ×10; c: ×12).

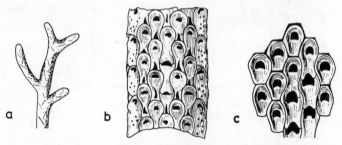

each other; pores present in the peripheral zooecia (so-called coscinozooecia). Aperture (opesium) semicircular with one pair of lower teeth. The zooecia are elongated to pear-shaped to hexagonal and arranged in longitudinal rows. Ovicells (ooecia) developed into a flat helmet shape on the distal portion of the skeleton (hyperstomial type). Occurrence: Cretaceous–Tertiary.

Two common Recent genera which are not calcified or only weakly calcified are:

Flustra (Fig. 187, a–c): Erect colony, bilamellar with narrow to wide lamellae; not calcified, flexible. Zooecia multiserial and enclosed by a pronounced ridge, more or less oval with a sickle-shaped aperture. The zooecia have distal spinae ('horned' zooecia). Pores present in the distal portion of the zooecium wall (septulae). Simple interzooecial avicularia (the zooecium is represented by an avicularium). Ovicell developed within the zooecium (endozooecial). Occurrence: Recent.

Bugula: Colonies bushy, unilamellar, chitinous and flexible. Zooecia more or less rectangular, arranged biserially or multiserially. Stalked avicularia on the lateral zooecia, shaped like a bird's head. Occurrence: Recent.

3.   Class: Phylactolaemata

Exclusively fresh-water bryozoans without calcareous hard parts. The lophophore is horseshoe-shaped. The zooecia are aggregated into clumps in a jelly-like fashion. The overwintering hard-shelled reproductive bodies are known as statoblasts. About 12 genera are known. Fossil forms have been identified with certainty since the Quaternary. One example is *Plumatella*. Occurrence: ? Tertiary–Recent.

The ecology of the Bryozoa

With a few exceptions, all bryozoans are sessile and stenohaline organisms and true facies fossils. They mostly inhabit shelf seas down to depths of 200 m, although their optimum living conditions are at depths between 10 and 80 m. In abyssal zones they have been found at depths of about 8200 m in the Kermadec trench.

Ecological factors such as the type of substrate, the water depth and water currents have a considerable influence over the morphology of the colonies, which can vary tremendously in different biotopes, even within the same species. This can greatly impede the systematic and taxonomic study of bryozoans. The extent to which environmental factors can influence the shape of bryozoan colonies is illustrated by a comparison of individual species from the basin facies and the marginal facies. Thus, the bryozoan assemblages of the Cretaceous chalk sedimented in calm water often have a completely different growth habit from the coastal sublittoral marginal facies of the Subhercynian in the northern Harz foreland. In the Santonian chalk facies the dendritic colonies were mostly slender and fragile, whereas in the Santonian marginal facies they were much more robust, sometimes forming crusty and tuberous, coarse fist-sized colonies. In this facies the tuberous bryozoan colonies are generally multilayered, while the dendritic cylindrical colonies are robust and solid.

Compared with the corals or calcareous algae, the bryozoans play a subordinate role in the formation of rocks. Even the well-known 'bryozoan' reefs of the Zechstein are mostly reefs made by the calcareous alga, *Stromaria schubarthi*. The large fan-shaped or cup-shaped 'fenestrate' (*Fenestella, Polypora*) or 'pinnate' (*Acanthocladia*) cryptostomes which had settled on these reefs only played a minor part as sediment trappers. True bryozoan sediments, which sometimes occur in the form of biostromes, and shallow-water faunas rich in bryozoans, are particularly well known from the tuffaceous chalk of France, Belgium, Holland and southern Sweden. Biostromes are flat laminar structures which are mainly composed of sessile organisms (e.g. bivalve, crinoid and coral banks and bryozoan mats). They are never mound- or lens-shaped. Bioherms, on the other hand, are reef-, mound- or lens-like structures, strictly organic in origin, which are embedded in a different type of rock.

15.   Phylum: Brachiopoda

The brachiopods are exclusively marine, bilaterally symmetrical organisms with a bivalved shell. In spite of their morphological and ecological similarities with the bivalves, there are some fundamental differences between the two groups. First, with respect to the living animal (i.e. anatomically speaking), brachiopod valves are at the 'top' and 'bottom', whereas in the bivalves they are 'right' and 'left'. However, the terms dorsal and ventral valves should be avoided because they are too easily related to the (biological) position of top and bottom which may well be different. Also, each brachiopod valve is bilaterally symmetrical, whereas in the bivalves the plane of symmetry runs between the two valves (in the plane of the commissure).

Most brachiopods attach themselves to a substrate by means of a fleshy stalk or pedicle. The valve from which the pedicle emanates is known as the pedicle valve. The opposed valve contains the two arms (hence the name brachiopod) or lophophores, which are usually propped up by a calcareous support. This valve is accordingly known as the brachial valve.

The name brachiopod was originally coined because it was thought that the 'arms' were in fact 'feet', as in the case of the bivalves. Because of their similarity with ancient Etruscan lamps, brachiopods are also commonly known as 'lamp-shells' (Fig. 191).

The brachiopods are divided into the two classes Inarticulata (without hinges) and Articulata (with hinges) (Table 7). They are dioecious.

Fig. 191. Organisation of an articulate brachiopod.

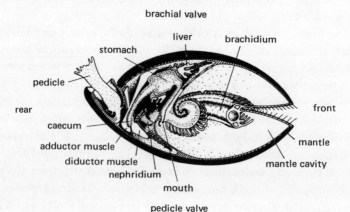

brachial valve

liver   brachidium

stomach

pedicle

rear   front

caecum

adductor muscle   mantle

diductor muscle   mantle cavity

nephridium

mouth   mantle

pedicle valve

The phylum reached its climax in the Palaeozoic period. Only about 70 Recent genera are known as opposed to about 1700 fossil genera (Fig. 192).

The soft body of the brachiopods consists of:

1. Two mantle lobes; the mantle epithelium secretes the calcitic or phosphatic-chitinous shell material.
2. The visceral sac which contains the alimentary canal with the mouth, intestine and liver, the muscles, and the heart, a contractile sac close to the stomach. The coelomic fluid acts as blood and is channelled through a branching system of vessels which sometimes leave visible impressions on the inside of the valve. These so-called pallial impressions may be of taxonomic importance;
3. Two flexible arms (lophophores) which are penetrated by a canal and often supported by a calcareous brachidium. They are furnished with tentacles which have the function of gathering food;
4. A fleshy muscular stalk (pedicle) which is covered by a 'horny' cuticle and serves to anchor the animal. The stalk emerges from between the two valves or from a pedicle foramen below the umbo of the pedicle valve. Brachiopods are fixed to the substrate by their stalk.

The opening and closing of the bivalved shell is effected by muscles. Three main groups of muscles are distinguished (Fig. 193):

1. adductor muscles which close the shell;
2. diductor muscles (divaricator muscles) which open the shell;
3. adjuster muscles (in articulate brachiopods) which adjust the shell relative to the pedicle.

Fig. 192. Geological distribution of the brachiopod orders.

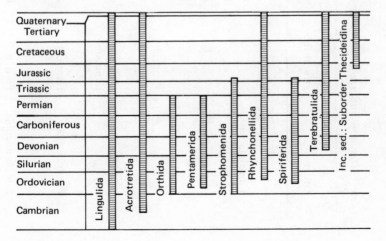

Table 7. *The most important diagnostic features of the brachiopods*

Order	Shell structure	Shell material	Pedicle foramen*	Brachidium	Hinge	Genera
Class Inarticulata						
Lingulida		'Horny'-phosphatic	Atrematous	Absent	Absent	*Lingula, Obolus*
Acrotretida		'Horny'-phosphatic	Neotrematous	Absent	Absent	*Orbiculoida, Discinisca, Crania*
Class Articulata						
Orthida	Impunctate, punctate	Calcareous	Protrematous	Brachiophores	Present	*Orthis, Platystrophia, Dalmanella*
Pentamerida	Impunctate	Calcareous	Protrematous with spondylium	Crura	Present	*Conchidium, Pentamerus*
Strophomenida	Pseudo-punctate	Calcareous	Protrematous, pedicle ± atrophied	Brachiophores	Present	*Strophomena, Leptaena, Rafinesquina, Chonetes, Productus, Oldhamina*

Pugnax

Spiriferida	Impunctate, punctate	Calcareous	Telotrematous	Helicopegmate (spirolophous)	Present	*Atrypa, Dayia, Uncites, Athyris, Cyrtia, Spirifer, Cyrtina, Spiriferina*
Terebratulida	Punctate	Calcareous	Telotrematous	Ancylopegmate (plectolophous) with or without median septum	Present	*Stringocephalus, Rensselaeria, Terebratula, Coenothyris, Terebratella, Zeilleria*
Inc. sedis Thecideidina	Punctate, fibrous	Calcareous	Pedicle absent, pedicle valve fixed	Two-lobed	Present	*Thecidea*

* The terminology is based primarily on the position of the pedicle opening:
Atrematous: no pedicle opening, at most a shallow furrow, the pedicle emerging from between the two valves, as in the case of the Lingulida.
Neotrematous: pedicle opening in the shape of a slit or hole in the pedicle valve of inarticulate brachiopods.
Protrematous: pedicle opening represented by an open delthyrium or closed by a deltidium.
Telotrematous: pedicle opening restricted by deltidial plates, as in the Terebratulida.

Fig. 193. Schematic representation of the opening and closing mechanism in an articulate brachiopod, showing the relative positions of the adductor muscles (white) and the diductor muscles (black).

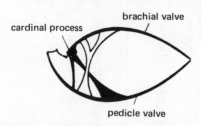

cardinal process

brachial valve

pedicle valve

In the Articulata the muscle apparatus is simple. In general, scars of the adductor and diductor muscles are found. Additional small scars are attributable to the adjustor muscles. Examples of the scars are found in the terebratulids and rhynchonellids.

In the hingeless brachiopods (Inarticulata) opening and closing the shell requires an especially complicated system of opposing muscles, and there are up to three additional pairs of muscles which move the two valves with respect to each other: the protractor and retractor muscles which permit sliding movements in the longitudinal direction, and the rotator muscles which permit rotating movements.

The shell

In the Inarticulata, the shell material secreted by the mantle epithelium consists of a chitinous-phosphatic material or of calcite. Thin layers of inorganic material often alternate between layers of organic material. Calcium phosphate is the main constituent. The shells of the Articulata are composed of calcitic material.

Composition of the shell material:

1. Periostracum: 'horny' organic layer.
2. Primary layer: fine fibrous layer on the outside of the shell. The longitudinal axes of the crystals are orientated at right angles to the shell surface with their C-axis. This layer is wholly inorganic and formed extracellularly.
3. Secondary layer: the calcitic fibres are oblique with respect to the primary layer. Each fibre is well defined and formed intracellularly. The C-axes are parallel to those of the primary layer and thus oblique with respect to the fibres of the secondary layer. All internal structures are formed from the secondary layer.

In the Articulata three types of shell structure are distinguished (Fig. 194):

1. Impunctate: the shell has no perforations; the most primitive shell structure.
2. Pseudopunctate: the shell is not perforated, but differs structurally from the impunctate type. It is penetrated by rod-shaped units (taleolae) and tends to weather more easily in those areas, giving rise to pore-like depressions. In the unweathered state, taleolae are represented by superficial tubercles on the inside of the shell.
3. Punctate: the valves are penetrated by canals running at right angles to the surface. The canals are in contact with the mantle and widen into trumpet-shaped cavities towards the outside. They may be arranged in regular rows or they may be irregular. Punctate shell structure was probably evolved repeatedly and independently in brachiopod phylogeny.

The structure of the shell is of taxonomic importance. The valves grow by the addition of material in a concentric pattern round the protegulum, the 'horny' larval shell. In many genera there is a depression along the median line of one of the valves, known as the sulcus. The opposed valve always has a corresponding saddle-shaped raised area, the fold.

The area between the umbo and the hinge line in one or both valves is often curved towards the opposed valve; this area is known as the palintrope. If the sculpture on the surface of the palintrope differs from that of the rest of the valve (the palintrope often has longitudinal or transverse striae), the palintrope is called the area or interarea.

An area which is flat in all directions is known as a planarea. An area curving in from a straight base (e.g. in the Spiriferida) is known as an interarea in the narrower sense. In many cases the rear margin is also the hinge line and of equal length in both valves. This shell type is known as a strophic shell with an interarea. In the English-language literature the terms area and interarea are generally equated.

In many Recent articulate brachiopods there are more or less star-shaped calcite bodies right under the mantle epithelium and free in the connective

Fig. 194. Shell structures of articulate brachiopods.

outside

impunctate   pseudopunctate   punctate

inside

tissue (designed to support the surrounding soft parts?). In the fossil state they have so far been found only in the Thecidea.

The relative position of the valves in articulate brachiopods is fixed by the hinge, which consists of two teeth in the pedicle valve and corresponding sockets in the brachial valve. The hinge teeth may be strengthened by plate-like dental supports. The spoon-shaped plate formed by converging dental plates and known as the spondylium serves as an attachment site for muscles in the palin-tropic area of the pedicle valve. A corresponding structure in the brachial valve of Pentamerida is called cruralium.

All structures close to the hinge line of the brachial valve are known as the cardinalia. These include parts of the hinge and the cardinal process, which is medially placed and which represents the attachment site for the dorsal ends of the diductor muscles. The cardinalia further include the supports for the arms, the brachiophores, which consist of two plates positioned either side of the pedicle foramen in the brachial valve of many orthids. They serve as the attachment sites for the fleshy lophophores. In the rhynchonellids, spiriferids and terebratulids the corresponding (homologous) structures are known as crura (singular crus). The crura form the proximal part of the calcareous arm supports in articulate brachiopods.

The lophophore support or brachidium consists of calcareous supports. Several important types of supports are distinguished, as summarised in Fig. 195.

The pedicle foramen and its protective plates are also of taxonomic significance. The terms used for the description of the pedicle opening and the surround-

Fig. 195. The most important types of brachiopod brachidia.

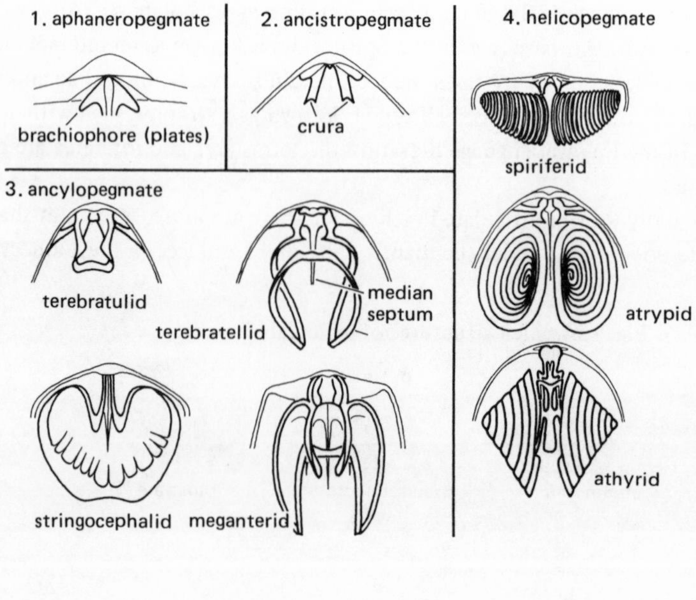

Fig. 196. The position of the brachiopod pedicle foramen (schematic representation): (1) epithyridid, (2) permesothyridid, (3) mesothyridid, (4) submesothyridid, (5) hypothyridid, (6) amphithyridid.

ing areas are given in Fig. 196. The triangular cavity under the umbo of the pedicle valve, and from which the pedicle emerges, is known as the delthyrium. The corresponding cavity in the brachial valve is known as the notothyrium. The delthyrium may be partly or completely closed by two calcareous deltidial plates forming the deltidium (Fig. 197). Correspondingly, the notothyrium may be closed by the calcareous chilidium. Deltidium and chilidium are here used as collective terms describing plates of different origins which close off the delthyrium and the notothyrium.

Fig. 197. Various degrees of development of the brachiopod deltidium (diagrammatic representation): (1) no deltidium, open delthyrium; (2) deltidial plates (deltarium, so-called deltidium discretum); (3) united deltidial plates (syndeltarium); (4) henidium; (5) xenidium.

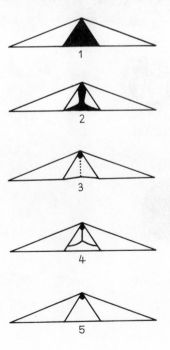

1.   Class: Inarticulata
        The shells of the Inarticulata consist of a 'horny' phosphatic material.
The shell material is calcareous only in the craniids. Inarticulata have no hinges or
lophophore supports. The pedicle opening may be atrematous, i.e. there is at
most a pedicle groove but no actual pedicle opening, or they may be neotrematous, if a special pedicle opening is present. The Inarticulata were a diverse
group with many species from the Cambrian to Silurian, but their importance
has declined sharply since then.

1.   *Order: Lingulida*
        Small elongated valves; long flexible pedicle. Occurrence: Cambrian-
Recent.
        Examples of the genera are *Lingula* (Fig. 198, 2), which is distributed world-
wide, has existed since the Ordovician and is one of the oldest genera ever, and

Fig. 198. Class Inarticulata (Order Lingulida): (1) *Obolus* (× 3); (2) *Lingula*
(× 1). (Order Acrotretida): (3) *Crania* (× 1); (4) *Orbiculoida* (× 1.2).

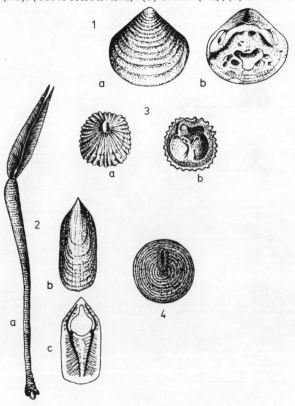

Obolus (Fig. 198, 1), known from the Cambrian to Ordovician of Europe, America and Asia.

2.   Order: *Acrotretida*
       Pedicle valve usually tall to flat conical shape; the brachial valve is flat. The outline is circular. The shell is 'horny' phosphatic, only occasionally calcareous. The pedicle opening is situated in the umbo of the pedicle valve or modified into a slit-like shape. Occurrence: Cambrian-Recent.
       Examples of genera are given by *Orbiculoida* (Fig. 198, 4; Ordovician-Cretaceous), *Discinisca* (Triassic-Recent) and *Crania* (Fig. 198, 3; Ordovician-Recent).
       The genus *Crania* has calcareous valves and is usually fixed to the substrate directly by its pedicle valve. The pedicle is lost during the course of ontogeny. Like *Lingula*, *Crania* is a genus which has survived over a long period of time and which existed with many species especially in the Ordovician and Upper Cretaceous (some index fossils). The name is derived from the fact that the sculpture on the inside of the pedicle valve looks like a skull (Latin, *cranium*).

2.   Class: Articulata
       This class embraces brachiopods with calcareous shells, a hinge, and in many cases lophophore supports. The Articulata have been in existence since the Cambrian and were particularly diverse from the Ordovician to the Upper Carboniferous (with a climax in the Devonian), in the Jurassic and the Upper Cretaceous. A number of species are important guide fossils.

1.   Order: *Orthida*
       Brachiopods with mostly biconvex valves and a straight hinge line, which were represented by a great number of species from the Cambrian to the Permian. Two suborders are distinguished according to the shell structure: (*a*) the Orthacea (impunctate); (*b*) the Dalmanellacea (punctate).
       In the case of the Orthacea, which were differentiated into many species in the Ordovician, two genera will be mentioned as representatives of certain morphological types which are nowadays split into numerous genera: *Orthis* and *Platystrophia* (Fig. 199, 1, 2). Both are common in the Lower Palaeozoic of the northern continents.

Fig. 199. Class Articulata (Order Orthida): (1) *Platystrophia* (× 0.7); (2) *Orthis* (× 0.7); (3) *Dalmanella* (× 1).

The Dalmanellacea, some of which are externally very similar to *Orthis* with their finely ribbed shells, include the genus *Dalmanella* for example (Fig. 199, 3): This genus was distributed worldwide from the Ordovician to Devonian.

The Orthacea and Dalmanellacea represent body plans which on the whole are not very variable and thus exhibit numerous homeomorphisms. They usually have long straight hinge lines, pronounced interareas and an open delthyrium.

2.   Order: Strophomenida

This order can be derived from the Orthida and, as in the case of the Orthida, the (straight) hinge line normally represents the widest part of the shell. All forms are pseudopunctate and usually have one concave and one convex valve with radiating ribs. The Strophomenida existed from the Ordovician to the Jurassic.

Important genera are:

Strophomena (Fig. 200, 1): Concave-convex (first-named is the brachial valve in each case) to resupinate, i.e. the valves change their curvature in the course of ontogeny; alternating fine and coarse ribs and concentric growth lines. The interarea has longitudinal striae. Abundant in the Ordovician of North America and the Baltic.

Fig. 200. Order Strophomenida: (1) *Strophomena* (× 0.7); (2) *Leptaena* (× 0.7); (3) *Rafinesquina* (× 0.7); (4) *Chonetes* (× 1).

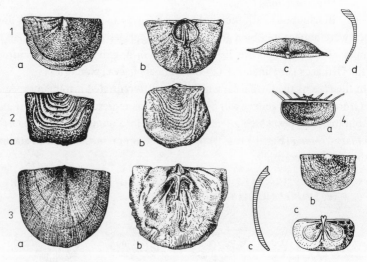

Leptaena (Fig. 200, 2): The shell shape is concave-convex, with pronounced trail. The valves have concentric wrinkly growth lines and fine radiating ribs. Occurrence: Ordovician–Carboniferous.

Rafinesquina (Fig. 200, 3): Concave-convex, as in *Leptaena*. Large forms with many species; Middle and Upper Ordovician particularly in North America, Europe and Asia. The interarea has longitudinal striae. Alternating fine and coarse ribs crossed by concentric growth lines.

Chonetes (Fig. 200, 4): Flattened concave-convex, semicircular valves. Deltidium (pseudodeltidium) present. Characteristic spines along the hinge line of the pedicle valve. The genus existed from the Middle Silurian to the Middle Permian, being particularly diverse in the Devonian. The chonetids are probably the ancestral group which gave rise to the Productacea.

Productus (Fig. 201, 1): Large forms with strongly convex pedicle valve; the brachial valve is concave. The valves are furnished with a variety of robust spines; some forms have trails.

Gigantoproductus (Fig. 201, 2): The largest known brachiopod from the Lower Carboniferous, with a 35 cm long hinge line. The productids were distributed worldwide in the Carboniferous and Permian.

Oldhamina (Fig. 201, 3): The shell shape of this aberrant group is concave-convex. The pedicle valve is hemispherically distended and has a median septum with up to 15 lateral septa. The brachial valve is deeply serrated into a feather-like shape. Interareas, pedicle openings and hinge teeth are absent. The genus *Oldhamina* is especially abundant in the Carboniferous and Permian of Asia (Permian of Timor). Forms with more (up to 35) lateral septa belong to the genus *Leptodus*.

Richthofenia (Fig. 201, 4): This genus is an index fossil for the marine Permian. It has very variable valves. The attached pedicle valve grows erect like a coral, and the uninhabited parts are divided off by transverse septa. On removal of the outer shell layer, a tall interarea with a deltidium becomes visible. The brachial valve is modified into a flat lid. Spines are developed ventrally and dorsally and are covered up by a secondary epidermis. The richthofeniids predominantly inhabited the warm oxygen-rich waters of the tropical belt of that time.

Fig. 201. Order Strophomenida: (1) *Productus* (×1); (2) *Giganto-productus* (×2); (3) *Oldhamina* (×0.7); (4) *Richthofenia* (×0.5).

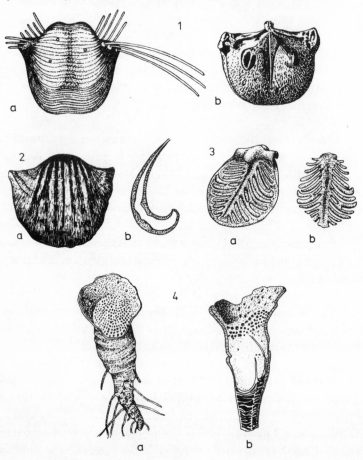

3.   Order: Pentamerida

This order is also directly descended from the orthids, and, like their ancestors, the Pentamerida have impunctate shells. The few representatives are large and strongly biconvex. They have a short hinge line and an open delthyrium. In accordance with the large size and strong curvature the cardinalia are well developed and form a spondylium in the pedicle valve. Occurrence: Ordovician–Permian.

Conchidium (Fig. 202, 1): Pronounced ribbing and a protruding beak. The spondylium is a deep and narrow spondylium simplex, i.e. supported by one (median) septum. Occurrence: Ordovician–Devonian.

Fig. 202. Order Pentamerida: (1) *Conchidium*, (a) transverse section of the umbonal region, (b) brachial valve with a spondylium (× 0.7); (2) *Pentamerus* (× 0.5).

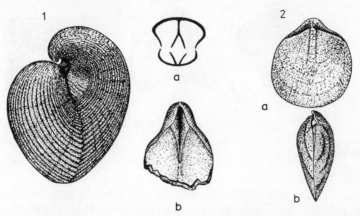

Pentamerus (Fig. 202, 2): Smooth or slightly crenulated; the long spondylium is supported by two parallel septa (spondylium duplex) so that the mould appears to be divided into five (from the Greek *penta*, five). Occurrence: Silurian.

4. Order: Rhynchonellida

The Rhynchonellida have impunctate and fibrous shells, which are more or less spherical in shape, and a short hinge line. They are usually furnished with a sulcus and a fold as well as a short protruding beak on the pedicle valve ('rostrate'). Simple biconvex crura form the base for the fleshy lophophore. The delthyrium is narrowed from the sides by deltidial plates, but a small opening usually remains under the umbo for the pedicle. Since the members of this order are all externally very similar, the septa and brachidia, which are very variable, have great taxonomic importance. Serial sections are necessary for accurate identification.

The Rhynchonellida are known to have existed since the Middle Ordovician; they are especially diverse in the Mesozoic.

Rhynchonella (Fig. 203): In the narrower sense this generic name applies to approximately triangular forms with pronounced ribs, a sulcus and a fold, dental supports and a dorsal median septum, which are widespread in the European Upper Jurassic (*Rh. loxiae*). Because of the difficulties involved in identifying individual species, the name '*Rhynchonella*' is used in the wider sense to include a number of forms of the Order Rhynchonellida. The Family Rhynchonellidae existed from the Triassic to the Cretaceous.

Fig. 203. Order Rhynchonellida: *Rhynchonella loxiae*, (a) brachial valve, (b) view from the front, (c) view from the side (× 2).

Pugnax: A mostly unribbed genus with a very prominent dorsal fold, occurring from the Devonian to the Permian.

5. Order: Spiriferida

This is a large and important group with predominantly impunctate shells, although a few genera also have punctate shells. The radially ribbed or folded shells are strongly biconvex. The spiral (helicopegmate) brachidium is characteristic. There is usually a well developed interarea in the pedicle valve but not in the brachial valve. The delthyrium of the pedicle valve is open or narrowed by deltidial plates. The order is common from the Middle Ordovician to the Upper Permian, with only a few forms surviving to the Lias.

The following are important genera with impunctate shell structure:

Atrypa (Fig. 204, 5): Pedicle valve usually flattened, with a sulcus; brachial valve curved more strongly. Both valves with radiating ribs. The brachidium consists of two simple spiral bands which initially bend outwards from the crura and then form a hollow cone with their tips towards the middle of the brachial valve. Occurrence: Silurian–Devonian.

Dayia (Fig. 204, 1): Small, smooth-shelled. The brachidial spirals are more or less planispiral and parallel to the median plane. The species *Dayia navicula* is a guide fossil for the Upper Silurian (Ludlow) of Europe.

Uncites (Fig. 204, 2): Relatively large forms with a long beak on the pedicle valve. The hinge line is short and curved. The deltidial plates meet and are strongly concave. The tips of the brachidial spirals point outwards. The genus is widespread in the Middle Devonian of Europe; guide fossil: *Uncites gryphus*.

Athyris (Fig. 204, 3): In this genus the two valves are equally curved with lamellar and sometimes spiny sculpturing. The tips of the brachidial spirals

Fig. 204. Order Spiriferida: (1) *Dayia* (×1); (2) *Uncites* (×0.5); (3) *Athyris* (×2), brachial valve; (4) *Spirifer* (×3); (5) *Atrypa* (×2).

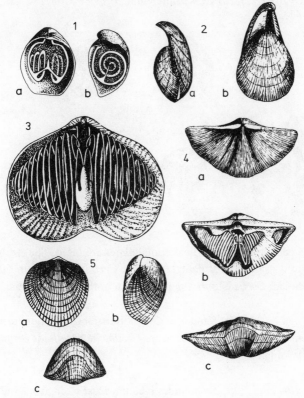

are directed outwards. The genus is distributed worldwide from the Devonian to the Permian (? Triassic) with the largest number of species in the Middle and Upper Devonian.

Spirifer (Fig. 204, 4): The original genus *Spirifer* is now split into a number of genera, so that the name is used only as a collective term here. The long straight hinge line is characteristic. The interarea of the pedicle valve is well developed, and the delthyrium is usually open. The pedicle valve has a pronounced sulcus and strongly diverging dental supports (easily recognisable on moulds). Moulds of the pedicle valve usually also show up clear muscle scars. Spiriferids form a very diverse and stratigraphically important group which is widely distributed in the Devonian and Carboniferous.

Cyrtia (Fig. 205, 1): The valves have a sulcus and a fold; they are impunctate and usually smooth. The pedicle valve is curved and has a very tall

Fig. 205. Order Spiriferida: (1) *Cyrtia* (×1); (2) *Spiriferina* (×0.5), (a) brachial valve, (b) front margin; (3) *Cyrtina*, (a) interior of the pedicle valve (×2), (b) view from behind, (c) view from the side (b, c: ×1).

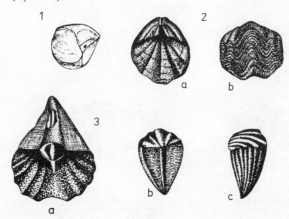

interarea with a pedicle foramen. The deltidial plates are fused. The strong curvature has given rise to robust dental supports. *Cyrtia* existed from the Silurian to Devonian; *Cyrtia exporrecta* is common in the Upper Silurian.

Cyrtina (Fig. 205, 3): Similar to *Cyrtia*, but ribbed and punctate. Widespread from the Silurian to the Permian. *Cyrtina heteroclita* is common in the Middle Devonian.

Spiriferina (Fig. 205, 2): Includes medium-sized strongly curved ribbed forms with punctate shells which have a prominent sulcus and fold. Represented with numerous species in the Triassic and Lias.

6. Order: Terebratulida

Punctate brachiopods, usually with smooth, occasionally with marginally folded valves; short hinge line and loop-like brachidium (ancylopegmate). Round foramen in the beak area. The order is presumably derived from the Dalmanellacea. The Terebratulida have been in existence since the Upper Silurian; they were particularly diverse in the Cretaceous.

Stringocephalus (Fig. 206, 2): Large biconvex forms, usually with smooth rounded valves. The umbo of the pedicle valve protrudes like a beak. There is a tall median ridge (septum) ventrally and a very long robust cardinal process dorsally. The genus is distributed worldwide in the Middle Devonian. *Stringocephalus burtini* is an index fossil for the Upper Middle Devonian (Givetian).

Fig. 206. Order Terebratulida: (1) *Rensselaeria*, (a) and (b) brachial valve (× 0.6), (c) hinge region from the inside (× 0.7); (2) *Stringocephalus*, (a) lateral view (× 0.3), (b) brachial valve (× 0.25), (c) hinge region from the inside (× 0.4).

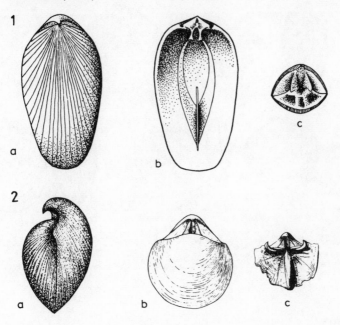

Rensselaeria (Fig. 206, 1): Relatively large oval biconvex shell with fine radiating ribs. The two loop-like brachidia fuse into a broad plate towards the front. The genus is relatively common in the Lower Devonian of Europe and North America.

Terebratula (Fig. 207, 3): Smooth, rounded to elongate oval valves with round pedicle foramen; rarely ribbed. On the frontal margin of the brachial valve there is an indentation bounded by two folds. The loop-like brachidium is short and has no median ridge. This is a former 'collective genus' which is now split up. The Family Terebratulidae has existed since the Upper Triassic; the Genus *Terebratula* in the narrower sense is restricted to the Miocene to Pliocene.

Coenothyris (Fig. 207, 4): This genus is smooth-shelled, usually elongate oval, rarely circular. The distal ends of the loop-like brachidium fuse to form a median plate; the brachidium is supported by a median septum. The genus is relatively common in the Triassic. In the Germanic Upper Muschelkalk (ceratite beds) *C. vulgaris* and *C. cycloides* are characteristic.

Fig. 207. Order Terebratulida: (1) *Terebratella* (x 0.5); (2) *Cincta* (x 0.7); (3) *Terebratula* (x 1); (4) *Coenothyris*, modifications of *C. vulgaris* (x 0.5).

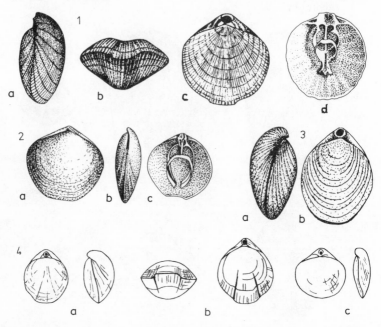

Terebratella (Fig. 207, 1): Shell smooth or with radiating ribs, with sulcus and fold as well as a low interarea. The loop-like brachidium is long and attached to the median septum by means of a transverse bridge. The genus has existed since the Oligocene, the Family Terebratellidae since the Lower Cretaceous.

Zeilleria: Smooth valves with a flat fold and sulcus. Two or more folds at the frontal margin. The loop-like brachidium is long and not attached to the dorsal median septum. Occurrence: Lower Jurassic–Upper Cretaceous.

Cincta (Fig. 207, 2): has a similar loop but is shaped like *Terebratula*. Occurrence: Lower and Middle Jurassic.

Incertae sedis: Thecideidina

The Thecideidina are a small group of uncertain status which may be strongly modified (neotenous?) Terebratulida. They are mostly small, up to 1 cm long, rounded to elongate forms with very unequal valves. The whole pedicle valve is normally attached to the substrate, and the pedicle itself is absent. The valves are punctate and fibrous. Dorsally there is a prominent cardinal process and wide hinge line from which the radial septa emanate. The Thecideidina have no brachidium. Their Recent representatives live in the warmer seas, e.g. the Genus *Lacazella. Thecidea* (Fig. 208), which is common in the Upper Cretaceous of Western Europe, is an important fossil genus.

Fig. 208. Thecideidina: *Thecidea.* (a) Interior of the brachial valve, (b) interior of the pedicle valve, (c) exterior of the brachial valve (× 2).

a   b   c

The biology and ecology of the brachiopods

1. Ecology. Virtually all present-day brachiopods are marine benthic sessile epifaunal filter-feeders. Among the fossil brachiopods only a few were possibly epiplanktonic; some became detritus-feeders. In the present-day forms attachment by the pedicle is permanent. The pedicle serves as both an anchor and a support. Hard substrates such as rocks, shells and corals, etc., are usually preferred as attachment sites.

Some Recent brachiopods are able to attach themselves to soft substrates such as algal stalks and ascidians, while others are able to settle directly on or in soft substrates. In *Terebratulina*, for example, the distal portion of the stalk is divided into fine rootlets which are able to penetrate calcareous substances, such as the *Globigerina* shells in the *Globigerina* muds, by means of acid secretions. Lingulids attach their pedicles straight into the mud with the aid of a sticky secretion.

Other forms such as the Acrotretida, Lingulida and others seem to have attached themselves to drifting plants since they are found in black shales without any bottom-dwelling fauna (epiplankton).

In all these cases the pedicle must have been sturdy and the pedicle foramen must have been large. There are also other types of attachment. Brachiopods with weak pedicles which do not support the shell but merely anchor it, tend to thicken their shell at the rear end of the valves in order to load and fix the position of the shell. In other brachiopods the pedicle may be totally absent, so that the shell is 'free' on the substrate with the rear end downwards.

In some Strophomenida the pedicle atrophied early in the course of ontogeny, usually in concave-convex forms which then lived unattached on soft substrates. If such brachiopod shells were turned over by currents, they were probably able to 'leap' up by rapidly opening and closing their valves and fall back into the 'right' position. Perhaps some were even able to swim like scallops (pectinids). 'Geniculate' forms in which the curvature of the valves changes during the course of ontogeny perhaps became 'sessile' again when the shell proved to be large and heavy enough, which is very probable in the case of the productids.

Direct attachment of the shell or one valve exists when the shell material secreted by the ventral mantle edge is in direct contact with the substrate. This is particularly common in biotopes with strong currents, e.g. in the case of present-day *Crania*. The site of attachment is usually restricted to a small area; this is particularly obvious in the richthofeniids. Secondary detachment can also occur after direct attachment, even in the case of the richthofeniids, as well as in the case of concave-convex shells such as the Genus *Oldhamina*, and in the Strophalosiacea with their numerous spines some of which probably functioned as roots. In the productids the dorsal spines may have clung to sediment, thus representing the transition to the infaunal lifestyle of which *Lingula* is an extreme. *Lingula* has been found embedded vertically in the sediment since the Ordovician.

2. Feeding. Brachiopods feed on suspended particles which are caught with the aid of the currents created by the beating cilia. Since there is no sorting process, inorganic particles are also included. Food consists mainly of diatoms and dinoflagellates. From the stomach the particles are sucked into the diverticula ('liver') and digested. Undigested material is returned to the stomach and channelled into the intestine, which contains a rotating mucous thread. Its main function is to form pellets, i.e. small coprolites. In the Inarticulata these are excreted through the anus, in the Articulata via the mouth. The coprolites are then transported to the mantle edge by ciliary action and eventually discarded by means of rapid opening and closing of the valves.

3. Ontogeny. Brachiopods are virtually always dioecious and sexually mature at an early stage (morphologically 'juvenile'). Fertilisation generally occurs in free water into which the genital products are shed. Some Articulata also have 'brood pouches' consisting of modified nephridia. A dense population is necessary for fertilisation, and brachiopods often occur in nests or localised clusters. Articulate larvae are only part of the free-swimming plankton for a few hours, inarticulate larvae for a few weeks. The latter are capable of developing much further ontogenetically and have even been found in the middle of the ocean.

Brachiopods grow throughout their lives, which extends over 7–8 years in *Waltonia*, for example. They reach sexual maturity after 2–3 years.

4. Enemies. Marine reptiles, ammonites (particularly dangerous in the case of juvenile brachiopods), fishes, crustaceans, starfishes and some boring gastropods are all possible predators. The shells of older brachiopods are often densely populated by foreign organisms such as sponges, hydrozoans, tube worms and occasionally bryozoans. Colonisation has the advantage of providing camouflage.

16.   Phylum: Conodontophorida

In 1856 Ch. H. Pander was the first to describe in detail tiny yellowy-brown tooth-like structures which he called conodonts. Conodonts measure 0.2–3.0 mm and consist of calcium phosphate which has the composition of a carbonate-apatite. Their status within the system of organisms is still unclear. Thanks to their wide distribution, their rapid evolution and their abundance of relatively easily identifiable forms, as well as their resistance to weathering, they are among the most reliable guide fossils from the Ordovician to the Triassic.

The division of the conodonts into three different types is very useful for practical purposes (Fig. 209):

1. 'Single teeth' or 'simple conodonts' (simple cones). The simple 'distacodide' form is found in the Lower Cambrian to Silurian.
2. 'Platform conodonts'. Above a very broad and rather flat 'lower part' (platform) there is a row of denticles or tubercles, the so-called carina, on the 'upper side'. The teeth are fairly even in size. The platform conodonts are found in the Ordovician to Triassic.
3. 'Compound conodonts'. Leaf- or branch-like conodonts with a narrow base. The teeth are very variable in size, and there is often one large main tooth in addition to lateral teeth of varying size. The position of the main tooth is an important taxonomic criterion. Compound conodonts are widespread from the Ordovician to the Triassic.

Part of the conodonts is a cone-shaped or plate-like basal body which is attached more or less loosely in the basal cavity, in the so-called pseudopulpa, on the underside of the conodont. The conodont and the basal body are collectively known as a holoconodont. However, only the conodont without the basal body is commonly found.

Internal structure and growth

Most conodonts are made up of lamellae. Conodonts with a fibrous structure are less common, but these probably primarily also had a lamellar

Fig. 209. Different types of conodont. (a) 'Compound conodonts': *Lingonodina* (left) and *Hibbardella* (right) (× 75), (b) 'Simple cones' or 'simple conodonts' (× 37). (c) 'Platform conodonts': left, oral (× 37); right, aboral (× 20) (*Palmatolepis*).

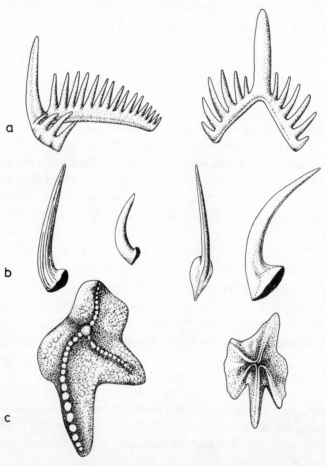

structure. In lamellar conodonts growth proceeds by the concentric addition of new external lamellae; growth is thus centrifugal. This is in direct contrast with the oral or dermal teeth of vertebrates which grow centripetally. The conodonts also show no recognisable functional wear, nor have a pulp-chamber like teeth do. They were able to regenerate broken-off pieces. This behaviour proves that they were surrounded by a tissue during growth (Fig. 210).

Studies of conodont ultrastructure have led to the establishment of three major groups, each of which represents a particular growth pattern (Bengtson, 1976): the protoconodonts, paraconodonts and euconodonts.

Fig. 210. Growth of a conodont, longitudinal section. The arrows
indicate the direction of growth and filling of the base.

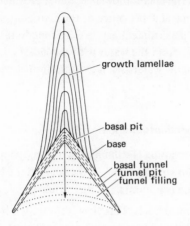

1. Protoconodonts (Upper Precambrian–Lower Ordovician): The oldest
group. Slightly curved simple cones with a deep basal cavity; cones of hyaline
phosphates. Growth was initiated at the tip; e.g. *Hertzina* (Fig. 212, a).

2. Paraconodonts (Middle Cambrian–Ordovician): Simple cones of hyaline
phosphates. Concentric growth centre just below the tip; few thick lamellae
growing in the basal direction; deep basal cavity. *Furnishina* (Fig. 212, b).
Broad flat area on the anterior side of the cusp (Upper Cambrian).

3. Euconodonts (Lower Ordovician–Upper Triassic): Consist of a conodont
proper and a distinct basal body, both constructed of lamellae. Concentric
lamellar growth. Euconodonts are divided into simple cones, bars, blades and
platforms.

The lamellar conodonts are sometimes preserved in their original arrangements,
which is a bilateral apparatus consisting of 14 to 22 pairs of conodonts. Finds of
conodont apparatuses and the reconstruction of multi-element apparatuses from
isolated individual elements permit certain conclusions about the shape of the
conodont-bearing animals, the conodontophorids. They were relatively small
(*c*. 2 cm) probably barrel-shaped animals with a planktonic lifestyle and may
have resembled present-day tunicates.

The ecology of the conodonts

Up until a few years ago the conodonts were thought to be exclusively
marine organisms distributed worldwide, because they were found in marine
sediments. However, new studies have revealed that there are differences between
conodont faunas from pelagic facies and shallow-water facies. Especially in the

Ordovician a biofacies-dependence (water temperature, salinity? etc.) of the conodont species seems to have led to the creation of large faunal provinces. The biofacies-dependence and geographical dependence of the conodontophorids can be explained in terms of a passive planktonic lifestyle. The conodontophorids probably filtered food particles from the water with the aid of a lophophore. Conodonts from extremely shallow waters seem to indicate that some forms also led a more or less benthic lifestyle.

The reconstruction of the Conodontophorida

The nature of the Conodontophorida is still unknown, and opinions on this problem differ widely. The following are some of the suggestions that have been made about the nature of the conodonts, i.e. they could be:

1. The internal supports of nektic animals. This view is supported by the capacity for regeneration and the lack of wear on the denticles.
2. Grasping tools of a digestive apparatus (not masticators).
3. Masticators from annelid worms of the polychaete group, whose fossil jaw apparatuses are known as scolecodonts. The conodonts (apatite) and the scolecodonts (chitinous-siliceous material) differ in their internal structure and the chemical composition of the denticles.
4. The apparatuses of jawless vertebrates (fish theory, visceral arcs) or small chordates with a barrel-shaped body, which according to their lifestyle may have resembled the tunicates.

Since the conodont apparatus was probably supported by a lophophore which surrounded the mouth of the animal, the systematic classification of the conodonts as part of the phyletic group of the Tentaculata is now supported (Fig. 211).

Taxonomy and biostratigraphy

Until recently conodont taxonomy was a purely artificial system since their classification was not based on the taxonomic entity of the species, but on so-called parataxa (artificial genera, etc.). Most conodont finds are isolated elements which were named parataxonomically and given the names of form genera. Rare finds of complete conodont apparatuses have revealed that several form genera constitute a single natural entity. Since then the aim has been to unite and name natural genera and species on the basis of complete apparatuses.

The conodonts have great practical value for the stratigraphy of the Palaeozoic in particular. Occurrence: Cambrian–Triassic.

Fig. 211. Hypothetical reconstruction of a conodont apparatus. The large arrows indicate the direction of water circulation, the small arrows the direction of ciliary movement.

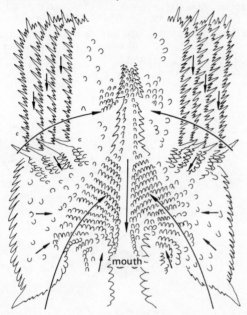

The following summarises some characteristic genera and their occurrence. Simple cone elements:

Drepanodus (Fig. 212, c): Laterally compressed with sharper anterior and posterior edges; cusp lenticular in cross-section. Occurrence: Lower Ordovician–Upper Devonian.

Compound elements:

Hindeodella (Fig. 212, e): Growth proceeded along two processes in particular, in which the posterior bar is longer, supporting different-sized denticles. Occurrence: Middle Ordovician–Middle Triassic.

Chirognathus (Fig. 212, d): Fibrous and hyaline with long thin denticles on the anterior bar; wide basal cavity. Occurrence: Middle Ordovician.

Polygnathus (Fig. 212, f): Leaf-shaped with platforms flanking the denticles; oral tubercles and ridges. Occurrence: Lower Devonian–Lower Carboniferous.

Fig. 212. Protoconodonts, paraconodonts, simple cone and compound euconodonts. (a) *Hertzina,* lateral view and cross-section (×53); (b) *Furnishina*, lateral and aboral views and cross-section (×53); (c) *Drepanodus*, lateral and anterior views and cross-section (×37); (d) *Chirognathus*, lateral view (×37); (e) *Hindeodella*, lateral view (×37); (f) *Polygnathus*, inner lateral, oral and aboral views (×37); (g) *Idiognathodus*, inner lateral and oral views (×38); (h) *Palmatolepis*, axial section through platform and basal body (×93); (i) *Amorphognathus,* oral view (×37).

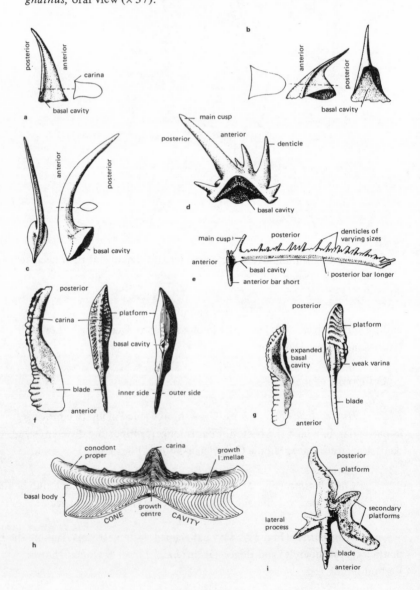

Amorphognathus (Fig. 212, i): Straight anterior and posterior processes have lateral, partly branched processes. Occurrence: Middle to Upper Ordovician.

Palmatolepis (Fig. 212, h): Sinuous carina with an outer and inner platform; inner platform often lobate with a lateral process. Occurrence: Upper Devonian.

Idiognathodus (Fig. 212, g): Leaf-shaped, expanded basal cavity; carina and blade are relatively straight; denticles replaced posteriorly by transverse ridges. Occurrence: Lower to Upper Carboniferous.

3.   Subgroup: Deuterostomia

Animals in which the larval mouth becomes the anus in early ontogeny.

17.   Phylum: Echinodermata

The Echinodermata are marine, stenohaline and benthic invertebrates with five-rayed symmetry. Their size varies between a few millimetres and *c*. 20 m. They have a water-vascular system, and a well developed skeleton consisting of calcareous plates or bodies, which in some forms is covered with flexible spines, bristles and sometimes also with pedicellariae (microscopic pincers). The skeleton is mesodermal in origin and secreted under the epithelium (internal skeleton!).

The plates consist of: (1) the internal stereome, (2) uniformly oriented crystalline calcite, and (3) the external organic stroma. The plates contain 3–15% magnesium carbonate, the proportion increasing with the water temperature. The microstructure of the calcareous plates is characterised by the presence of cavities which are recognisable even in the smallest fragments. During the process of fossilisation each plate gives rise to a crystallographically uniformly oriented calcite crystal. The original microstructure is generally still recognisable in thin sections. The stroma is replaced by secondary calcite in the process of fossilisation.

The diverse phylum of the Echinodermata used to be divided into the stalked or otherwise attached Pelmatozoa and the mobile Eleutherozoa. However, new knowledge led to the abolition of this division, and instead the phylum is now divided into five subphyla:

1. Homalozoa (carpoids)
2. Blastozoa (cystoids and blastoids)
3. Crinozoa (sea lilies, etc.)

4. Asterozoa (starfishes, brittle-stars)

5. Echinozoa (sea-urchins and sea-cucumbers)

The larvae of the Recent echinoderms can be related back to the hypothetical ciliated bilaterally symmetrical dipleurula larva. The primary bilateral symmetry of the body plan is disguised by secondary pentameral symmetry in the course of ontogeny and evolutionary history.

Occurrence: The echinoderms first appeared in the Cambrian. Although they were still represented by predominantly sessile forms in the Palaeozoic, the Echinoidea (sea-urchins) and other freely mobile forms such as the Asterozoa (starfishes and brittle-stars) and Holothuria (sea-cucumbers) gradually gained importance throughout the Mesozoic and Cenozoic. These days the mobile Echinodermata predominate by far (Fig. 213).

Morphology

From the central mouth a long convoluted alimentary canal leads to the anus which is either displaced slightly sideways or present on the opposite side. The gut is surrounded by the secondary body cavity (coelom) containing parts of the nervous system, the reproductive organs and the ambulacral system (water-vascular system). The water-vascular system is in communication with the outside water via the pores of the madreporite (sieve plate) or osmotically through the walls of the ambulacral tentacles or tube-feet. In the sea-urchins

Fig. 213. Stratigraphic occurrence of the Subphylum Homalozoa and of various echinoderm classes.

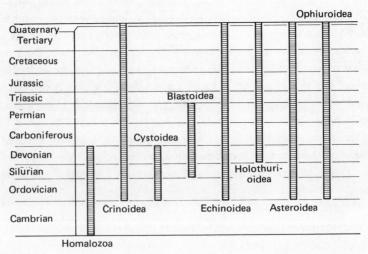

and starfishes the madreporite is linked to the stone canal which owes its name
to the extensive calcification of its walls in the starfishes. The stone canal joins
the circum-oral ring which surrounds the mouth and gives rise to five larger
radial canals. These radiate out in the direction of the ambulacra and send out
expandable tube-feet (or tentacles) to the outside through pores in the
ambulacral plates. The tube-feet serve the purposes of locomotion, respiration
or feeding. Even the Edrioasteroidea apparently had a well developed ambulacral
system in the Lower Cambrian.

Ecology

With the exception of a few brackish-water holothurians, the echinoderms are exclusively marine and occur down to great depths. Some live on
muddy substrates, e.g. some sea-urchins and holothurians, while others live on
sandy or rocky substrates or burrow in the sediment. Present-day crinoids live
on the sea-bottom, attached by a stalk, although they are more commonly
free-swimming. The Homalozoa, Crinozoa and Blastozoa have microphagous
feeding habits.

1.   *Subphylum: Homalozoa (Carpoidea)*

Asymmetric flattened forms which consist of a theca and a stalk-like
process. In the Class Stylophora the latter is known as the aulacophore and
interpreted as an ambulacrum-carrying arm, although it was originally thought
to be a stalk. The Homalozoa have two different sides and probably lay flat
on the substrate. Little is known about the positions of the mouth and anus.
The sculpturing is variable and some forms are covered with spines. Various
pores perhaps served as an aid to respiration. The Homalozoa represent
a relatively diverse subphylum which was restricted to the Lower Palaeozoic
(Cambrian–Devonian) and seems to have become extinct without leaving any
descendants. Because of their asymmetry they form a very isolated group
within the Echinodermata.

Ubaghs and Caster (Moore & Teichert *Treatise on Invertebrate Paleontology*;
Part S, 1967) distinguished the following classes:

Homostelea (Cambrian): *Gyrocystis*
Homoiostelea (Cambrian–Devonian): *Dendrocystites*
Stylophora (Cambrian–Devonian): *Cothurnocystis*

In spite of their undoubted echinoderm affinities, the Stylophora are considered
by some authors (Jefferies, 1979) to be chordates and are thus called Calcichor-
data. According to Jefferies, they were ancestral to the Tunicata, as well as to

Fig. 214. Subphylum Homalozoa: (1) *Dendrocystites* (×0.5);
(2) *Cothurnocystis* (×0.6); (3) incertae sedis: *Cymbionites* (×1).

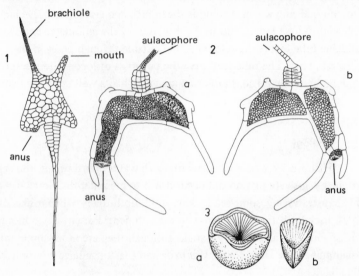

the Acrania and the Vertebrata. In this interpretation the aulacophore is the tail,
and it accordingly forms the rear part of the animal.

Examples: *Cothurnocystis* and *Dendrocystites*, both from the Ordovician
(Fig. 214, 1, 2).

The so-called Haplozoa represented in the Australian Lower Cambrian by the
genera *Cymbionites* (Fig. 214, 3) and *Peridionites* are of uncertain status.

2.   Subphylum: Blastozoa

This subphylum embraces the Palaeozoic classes of the former
'Pelmatozoa'. All are furnished with brachioles (short unbranched arms). The
following classes are included: (1) Eocrinoidea, (2) Parablastoidea, (3) Blastoidea,
(4) Cystoidea. Only the well represented classes of the Blastoidea and Cystoidea
will be characterised briefly here.

1.   Class: Blastoidea

The blastoids have pronounced pentameral symmetry and are stalked.
In the younger, more specialised forms the relatively small, more or less bud-
shaped theca consists of 13 firmly connected plates. In detail, these comprise:
five deltoid plates (interradials, surrounding the mouth), five radials (deeply
indented) and three basals. The mouth and anus are covered by small plates.
The five ambulacra are arranged in longish 'petals' and are furnished with

numerous thin brachioles. Along the median line of each ambulacrum there
is a lancet plate. The so-called hydrospires, infolded thin-walled respiratory
structures, lie below the ambulacra. The folds of the hydrospires unite to form
a longitudinal canal which terminates near the mouth as a special opening, the
spiracle. The spiracles were presumable used for respiration. In the blastoids the
thecal plates have no perforations equivalent to the diplopores in the cystoids.

The blastoids were exclusively marine and widespread from the Silurian
to the Permian. They are particularly diverse and well preserved in the Lower
Carboniferous limestones of North America. They are also found in the Permian
of the Island of Timor.

Timoroblastus (Fig. 215, 1): Star-shaped theca with short sunken
petal-shaped ambulacral areas. Occurrence: Upper Permian of Timor.

Schizoblastus (Fig. 215, 2): Bud-shaped to spherical theca. The
ambulacral areas extend over the entire theca. In each ambulacral area there
are up to four hydrospires. Occurrence: Lower Carboniferous (Mississippian) of
North America and common in the Permian of Timor.

Pentremites (Fig. 215, 3): Bud-shaped to spherical theca with convex
base. Wide petal-shaped ambulacral areas. Two spiracles and the anus are united
into one large opening. Occurrence: Very common and well preserved in the
Lower and Upper Carboniferous (Miss. and Pennsylvanian) of North America.

2.   Class: Cystoidea
The cystoids were widespread in the Lower Palaeozoic. Their spherical
or sac-like thecae consisted of numerous (up to 2000) irregularly arranged
plates. The presence of biserial and unbranched brachioles without pinnules
and the pore structure of the plates are characteristic. Two pore types are
distinguished: twin pores (diplopores) which are arranged in pairs and irregularly
distributed over the surface, and pore-rhombs (dichopores) which are arranged
in various rhomboidal patterns (Fig. 216). The comb-shaped pectinirhombs are
specially differentiated dichopores which are developed only on a few plates.

The mouth is situated at the apex of the theca, and the anus is adjacent
or displaced to the side. The genital pore and the hydropore are generally
between these two openings. There are two to five short ambulacra.

. The two orders Rhombifera and Diploporita which are distinguished on the
basis of the two pore types were clearly separated from the very beginning
(Lower Ordovician). They have lately been regarded as classes in their own
right.

Fig. 215. Subphylum Blastozoa (Class Blastoidea): (1) *Timoroblastus*,
(a) aboral, (b) lateral, (c) oral view (×1); (2) *Schizoblastus*, (a) lateral,
(b) oral view (×1); (3) *Pentremites*, (a: ×1; b: ×0.6). A, ambulacrum;
B, basal plate; D, deltoid plate; M, mouth; R, radial plate; Sp, spiracle.

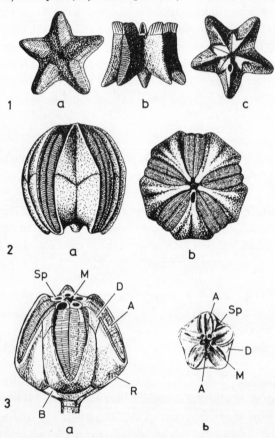

Fig. 216. Arrangement of the pores and pore canals in the Cystoidea.
(1) Diplopores, viewed from above and in section. (2) Dichopores:
(a) conjunct pore-rhombs, (b) disjunct pore-rhombs (pectinirhombs),
both epithecal; (c) (hypothecal) pore-rhombs not visible from outside;
(d) section.

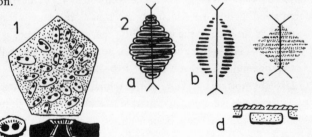

1.   Order: Rhombifera

With pore-rhombs. The pores are depressions or tubes in adjacent plates at right angles to the plate margins. They may be completely open to the outside or only open at the end. Occurrence: Ordovician–Devonian.

Echinosphaerites (Fig. 217, 1): With over 300 individual plates. Smooth in the unweathered state; the pores are visible only in weathered specimens. All thecal plates have pore-rhombs. The theca is 2–4 cm in diameter. Occurrence: Ordovician.

Fig. 217. Class Cystoidea: (1) *Echinosphaerites* (A, anus; G, gonopore: × 0.5); (2) *Sphaeronites* (× 1).

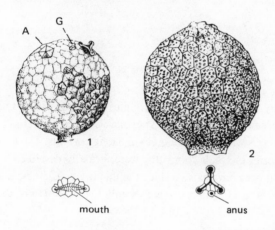

Pleurocystites (Fig. 218): Theca flattened laterally. One side consists of 24 thecal plates. The other side has many additional plates and a large anal orifice close to the base of the stalk. Pectinirhombs. Large brachioles, a hydropore and a gonopore next to the mouth. Long flexible stalk. Occurrence: Ordovician.

2.   Order: Diploporita

Numerous small plates. The pores extend vertically through the thecal wall. Pore openings generally paired. Ambulacra commonly present on the calyx. Occurrence: Ordovician.

Sphaeronites (Fig. 217, 2): Spherical or oval cystoids. Oral area with three to five radiating ambulacra. Occurrence: Middle Ordovician.

Fig. 218. Class Cystoidea: *Pleurocystites* (×1). P, pectinirhomb.

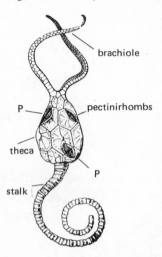

brachiole

P — pectinirhombs

theca

P

stalk

3.   Subphylum: Crinozoa

Predominantly sessile echinoderms with mostly pentameral symmetry, attached by a stalk or the base of the calyx. They have a calyx with arms (brachia) directed upwards which catch food by creating currents. The crinozoans include the majority of sessile forms that were formerly described as pelmatozoans. The Crinozoa have existed since the Cambrian. They include the two classes Paracrinoidea and Crinoidea of which only the latter will be described.

1.   Class: Crinoidea (sea-lilies)

Predominantly stalked, sessile-benthic, rarely free-swimming, highly differentiated crinozoans. They consist of a segmented stalk (columna), a cup (calyx, theca) with a flexible roof (tegmen) and arms (brachia). The calyx and arms are collectively known as the crown. The calyx consists of regularly arranged rings of calcareous plates (Fig. 219).

The upper side of the calyx is the ventral (oral, actinal) side, while the underside is the dorsal (aboral, abactinal) side.

The ambulacral system is concentrated in the arms. The calyx envelops the soft parts; its dorsal part may continue as the stalk.

The base of the calyx may be monocyclic or dicyclic (Fig. 220). A monocyclic base has only one ring of five basals positioned interradially, i.e. in the extended interbrachial areas. Dicyclic bases have two rings of plates. The plates under the basals are radially positioned in line with the arms; they are known as the infrabasals. The infrabasals are often quite inconspicuous. Above the

Fig. 219. Organisation of a crinoid (schematic).

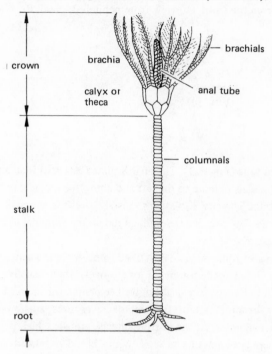

Fig. 220. Organisation of the thecal base in the crinoids. B, basal; Br, brachial; Ib, infrabasal; R, radial.

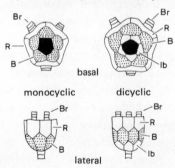

basals the radials are arranged in direct line with the arms.. The radials are followed by the brachials (brachial plates) of which the lowermost may be incorporated into the calyx.

The (posterior) anal interradial is specially differentiated. A special anal plate ('anal X' in the Inadunata and Flexibilia, 'tergal' or 'primanal' in the Camerata) is developed which may be accompanied by a series of additional

Fig. 221. Arrangement of the thecal plates in the Inadunata and Flexibilia: (a) view from the front, (b) view from behind. B, basal, Br, brachial; R, radial; RA, radianal; X, anal X.

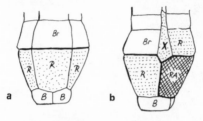

plates and a so-called anal tube (Fig. 221). The anal X plate (basianal) is an anal plate between the dorsal radials oblique to the left and above the radianal in the Inadunata and Flexibilia. The anal X plate may also lie over the aniradial, as in the case of the Disparida. The radianal lies under a radial; the aniradial lies over a radial or takes its place.

At the base of the calyx of some dicyclic unstalked forms there is a centro-dorsal plate. The calyx is closed at the top by a roof (tegmen) which usually consists of a leathery skin in Recent forms and of well cemented calcareous oral plates in fossil forms. From the mouth five ambulacral grooves radiate out towards and into the brachial bases. At the bottom of each groove, which is lined with epithelium, there is an ambulacral vessel with a blood vessel and nerve cord above. The ambulacral vessels unite to form a circum-oral ring (ring canal) from which short canals open into the body cavity and provide the ambulacral system with water that enters through pores in the oral disc. Each segment of the stem ('Trochit') is penetrated by an axial canal which is in communication with the chambered organ in the lower part of the calyx. The latter consists of five radially arranged chambers which are in communication with nerves and gonads in the arms. Crinoids have no eye spots or sense organs, but they have numerous tactile sense cells so that they are sensitive to touch.

The food current runs along deep grooves in the arms and pinnules which are lined with epithelium. Crinoids feed on microorganisms such as diatoms, Infusoria, larvae, etc. The arms are articulated with the radials; they are continuations of the ambulacral grooves on the tegmen. The number of plates (brachials) making up the arms can be in the order of several hundred thousand.

Depending on the arrangement of the brachials, uniserial, alternating and biserial arms are distinguished. The arms may be unbranched or extensively branched. The branched portions of the arms may all be equal (isotomous) or unequal in length (heterotomous). Only arms with isotomous branching usually have two rows of pinnules on the upwardly directed sides of the brachials. The pinnules also serve to collect food. In the Recent crinoids the

Fig. 222. The division of the arms in the Pentacrinoidea: (A) atomous, (B) isotomous, (C) heterotomous, (D) metatomous, (E) endotomous, (F) paratomous, (G) holotomous. In (E)–(G) the lateral branches are called ramuli, in (D) ramiculi. (G_1) shows a monostichous, (G_2) a distichous arrangement of brachials in the arms.

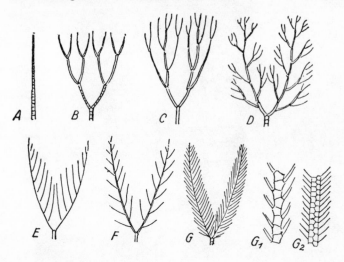

pinnules contain the sexual organs (Fig. 222). The type of branching, the absence or presence of pinnules and the connections between the brachials (arm segments) are all of great taxonomic importance. There are movable connections between the brachials (synarthry) and immovable ones (synostosis), and the articulating surfaces are accordingly differently developed. If the immovable articulating surfaces have ribs (crenellae) and grooves running radially from the central nerve canal, so-called syzygy is present. True fusion of the articulating surfaces is rare and is known as ankylosis.

The stalk can grow up to several metres in length (e.g. in *Seirocrinus c.* 18 m) or be completely reduced so that the crown itself is directly attached to the substrate (e.g. *Cyathidium*) or be free-living (e.g. *Marsupites* and *Antedon*). The stalk segments (columnals) are circular, oval pentagonal or star-shaped in cross-section with radiating ribs (crenellae), grooves or small pits on the surfaces of contact, which act as attachment sites for connective tissue fibres (Fig. 223).

The stalk may bear whorls of cirri which serve to anchor the animal and also have a respiratory function. In Recent crinoids they are continuously in motion. Columnals with cirri are larger and wider (nodes) than the others (internodes). The number of nodes increases towards the crown. The crinoid stalk is commonly attached to the substrate by root-like processes or flattened adhesive discs. In free-living forms grasping organs may be developed for temporary attachment.

Fig. 223. Thecal and stem elements with articulating surfaces from crinoids of the Subclass Articulata: (1) and (2) *Balanocrinus* (×1.5); (3) *Seirocrinus* (×2); (4) and (5) *Pentacrinus* (×1.5); (6) *Isocrinus* (×2); (7) *Bourgueticrinus* (×0.7); (8) *Austinocrinus* (left, ×1; right, ×0.7), (9) *Isselicrinus* (×1.2); (10) and (11) *Nielsenicrinus* (×1).

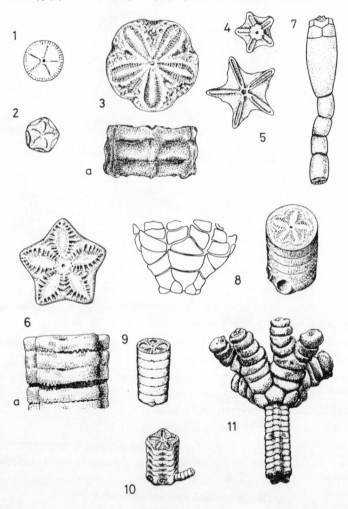

The spherical balloonings of the end of the stalk (lobolites) in certain crinoids (*Scyphocrinites*) are regarded as buoyancy organs.

The size of the crinoids is very variable. Some forms measure only a few millimetres, while others may grow up to 20 m in length including the stalk. In the Upper Cretaceous of North America a *Uintacrinus* calyx was found which had arms exceeding 1 m in length. However, on average, crinoid crowns are about 5–10 cm long.

The Crinoidea reached their prime in the Palaeozoic (Silurian to Lower Carboniferous). At that time they preferred marine habitats close to the coast. Of the 630 or so surviving species 80 are residual forms of the deeper waters down to depths of 180–1000 m. The other 550 Recent species are unstalked, and many of them live in the littoral zone. Like the Palaeozoic species, Recent crinoids also live in large lawns or assemblages.

1.   Subclass: Inadunata

The Inadunata (from the Latin *inadunatus*, not united: i.e. the arms are free above the radials) have monocyclic or dicyclic bases. The thecal plates are firmly connected; the mouth is covered by the tegmen. The Inadunata generally have anal plates ('anal X'), but no interradials. They are represented by about 1750 species from the Ordovician to the Triassic.

Examples:

Cupressocrinus (Fig. 224, 1): Fairly large, shallow, bowl-shaped calyx consisting of five basals, five radials and a large dorsocentral plate (i.e. dicyclic);

Fig. 224. Subclass Inadunata: (1) *Cupressocrinus* (× 0.5); (2) *Cyathocrinites* (× 0.5). B, basal; IB, infrabasal; R, radial; RA, radianal.

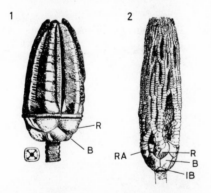

there are no interradials. At the upper margin of the calyx there is a 'consolidation apparatus' (oral plates) which is very specialised and serves as the attachment site for the arm muscles. The arms are broad, thick and undivided; the stalk is rectangular in cross-section and has one central and four peripheral canals. Occurrence: Middle Devonian.

Cyathocrinites (Fig. 224, 2): Dicyclic base; at the top of the radials there are small horseshoe-shaped articulating surfaces. With anal tube and anal

X; arms with isotomous branching. Stalk round in cross-section. Occurrence: Silurian – Carboniferous.

Timorocrinus: With long anal tube which forms constrictions in order to protect the arms. Occurrence: Permian of Timor.

2.   Subclass: Flexibilia
The Flexibilia are probably derived from the dicyclic Inadunata. They have dicyclic bases, and the lower row of plates consists of one small and two large infrabasals. Articulation between the thecal plates is possible (articulating surfaces). The mouth and ambulacra are visible. The arms are uniserial without pinnules and often roll up towards the inside. Anal tube only in the Order Taxocrinida. The lower brachials are incorporated into the calyx. The stalk is round in cross-section and has no cirri. About 300 species. Occurrence: Middle Ordovician–Upper Permian.

Taxocrinus (Fig. 225): Between the radials there is a large interradial which is followed by more interradials. Anal tube present. The arms were branched in a complicated manner. Occurrence: Lower Devonian–Lower Carboniferous.

Fig. 225. Subclass Flexibilia: *Taxocrinus* (×1.5). A, anal tube; Ax, anal X.

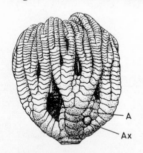

3.   Subclass: Camerata
The Camerata have monocyclic or dicyclic bases; all thecal plates are firmly connected. The uniserial or biserial arms bear pinnules and branch at least once. The mouth and ambulacral grooves are concealed under the tegmen. The tegmen is variable: with anal tube, and thorns projecting laterally, etc. Interradials always present in the anal interradius. (Proximal interradial in the anal sector = tergal.) Occurrence: Middle Ordovician–Upper Permian. About 2500 species; very diverse in the Lower Carboniferous.

Examples:

Hexacrinites (Fig. 226): Very simple calyx with monocyclic base. Only three basals, five large and very tall radials and one interradial (tergal) resembling the radials. The tegmen is slightly arched; the columnals are round and furnished with spines. Common in the Devonian of Europe; rare in North America.

Fig. 226. Subclass Camerata: *Hexacrinites.* (a) Calyx viewed from the side; (b) plate diagram; (c) columnal (x 1).

 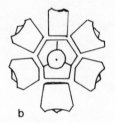

a   b   c

Platycrinites (Fig. 227, 2): Monocyclic base; three basals of unequal size, five large tall radials. The interradials are displaced to the tegmen. Numerous arms, occasionally branched distally; well developed pinnules. The thecal plates are often ornate. The stalk is twisted like a cork-screw, and the columnals are oval in cross-section. Occurrence: Devonian–Permian; very diverse in the Lower Carboniferous.

Scyphocrinites (Fig. 227, 1): Large, elongated pear-shaped calyx. The height of the crown can reach about 70 cm. Four basals, numerous interradials, interbrachials and interpinnulars make up the calyx which consists of a particularly large number of individual elements. In each radius there are two isotomous arms. The stalk is round and consists of flat columnals. The articulating surfaces of the columnals have crenellae and are perforated by a pentameral star-shaped canal. There are spherical lobolites on the roots of *Scyphocrinites* which are probably buoyancy organs. Occurrence: Silurian–Devonian.

Gilbertsocrinus (Fig. 227, 3): The basals and radials form independent rows of plates. The radials are separated by interradials. The tegmen has interradial, plated, tube-shaped processes forked at the end and directed downwards. Round stalk. Sometimes in association with the parasitic snail *Platyceras*. Occurrence: Middle Devonian–Lower Carboniferous.

Fig. 227. Subclass Camerata: (1) *Scyphocrinites.* Calyx (×0.5);
(a) articulating surface of a columnal; (b) longitudinal section of several
columnals (a, b: ×1.2); (c) lobolite (×0.5). (2) *Platycrinites* (×1).
(3) *Gilbertsocrinus* (×0.3). B, basal; R, radial.

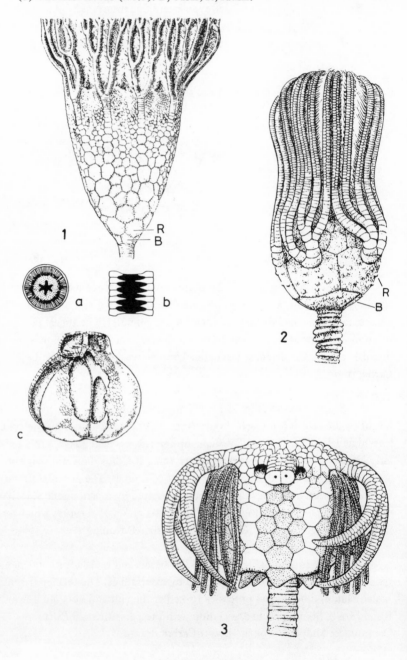

4.   Subclass: Articulata

This subclass embraces nearly all Mesozoic and younger crinoids. The Articulata are highly differentiated and have a dicyclic or pseudo-monocyclic base. The infrabasals and sometimes the basals too are atrophied or completely absent. The calyx is generally small, and the mouth and ambulacral grooves are free. The arms are uniserial, occasionally biserial and always furnished with pinnules. Anal plates absent. Many unstalked forms.

The Articulata have existed since the Lower Triassic and were most diverse in the Malm. They are still very diverse at the present time.

Examples:

Encrinus (Fig. 229, 1): Shallow calyx with a dicyclic base. The five large radials are trapezoid; the five infrabasals are very small and hidden under the uppermost columnal (Fig. 228). *Encrinus liliiformis* has ten, *E. carnalli* 20 biserial arms close together. The stalks are round, without cirri and attached to the substrate by a thickened disc. The columnals have crenulated margins. Occurrence: Germanic Muschelkalk (Triassic). The 'Trochiten' limestone consists of many isolated columnals of *E. liliiformis. Encrinus* is placed by some authors among the Inadunata.

Fig. 228. Thecal base of *Encrinus liliiformis* (×1).

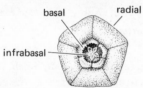

Seirocrinus (Fig. 223, 3): Very long stalk with nodes closely packed at the proximal end. Large radials; proximal pinnules reduced. Arms with weak isotomous branching. Short cirri. The articulating surfaces of the columnals have a star-shaped or pentameral pattern with serrations. Occurrence: Lias–Dogger.

This genus includes *Seirocrinus subangularis* from the Lias (*Posidonia* shales of Holzmaden) of which crowns have been found with up to 1400 branches. Stalk lengths of up to 18 m have been recorded.

Pentacrinus (Fig. 223, 4, 5): Similar but has a short stalk, heterotomously branched arms and long cirri. The columnals of this genus are also pentameral. Occurrence: Triassic–Malm.

Both genera are represented by many species in the Lias.

Fig. 229. Subclass Articulata: (1) *Encrinus* (×0.8); (2) *Apiocrinus* (×0.5, ×1); (3) *Seirocrinus* (×0.6, ×1). B, basal; R, radial.

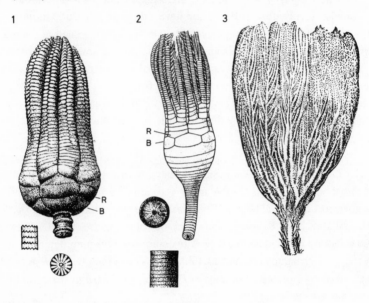

Apiocrinus (Fig. 229, 2): Pear-shaped calyx with a gradual transition to the proximally thickened stalk. The uppermost columnal has five radial edges on its upper side which slope down like a roof. Above, there are five basals and three rows of five radials each. Occurrence: Jurassic.

Bourgueticrinus: This genus in the Cretaceous is similar. Small, squat, with a monocyclic base and short stalk. The articulating surfaces of the columnals have a transverse ridge.

Marsupites (Fig. 230, 2): Spherical calyx made up of large radially sculpted plates. Unstalked. Centrodorsal plate large and pentagonal. Five infrabasals, five basals and five radials. The radials have a horseshoe-shaped articulating surface for the arms. Occurrence: Upper Cretaceous (Upper Santonian; *Marsupites* beds).

Uintacrinus (Fig. 230, 1): Spherical theca consisting of numerous small unsculpted plates. Centrodorsal plate small and pentagonal. Five small basals, five radials and numerous interradials and interbrachials. Ten undivided arms (over 1 m long). Unstalked, presumably free-swimming. Very common in the Upper Cretaceous of Kansas, for example.

Fig. 230. Subclass Articulata: (1) *Üintacrinus* (× 2); (2) *Marsupites* (× 0.8); (3) and (4) *Saccocoma*, lateral and dorsal view (× 5). B, basal; C, centrodorsal; Ib, interbasal; Ifb, infrabasal; Ir, interradial; R, radial.

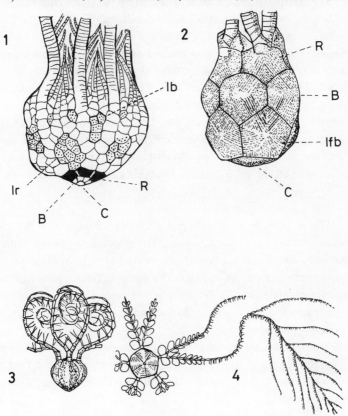

Saccocoma (Fig. 230, 3, 4): Small unstalked free-swimming crinoids with five long arms, each divided into two, which can roll up spirally. Calyx consisting of five radials and a very small central plate. The ends of the arms have many lateral branches. The proximal brachials are furnished with leaf-like lateral processes. Occurrence: Upper Jurassic–Upper Cretaceous. Very common in the Solnhofen limestone (Malm ς).

Antedon: Recent articulate crinoid with ten arms arranged in pairs round the mouth. On the underside of the calyx there are about 40 cirri which anchor the animal to the substrate. Locomotion is effected by the arms and cirri. No active swimming but rather passive drifting, or hemisessile on the sea-bottom.

4.   Subphylum: Asterozoa

1.   Class: Stelleroidea

The Asterozoa, which are represented by the single class of the Stelleroidea, are free-living animals usually consisting of a flat central disc and five radiating arms. The mouth opening is in the middle of the underside. The ambulacral appendages are also directed downwards and serve the purpose of locomotion. The water-vascular system radiates from the mouth into the tip of each arm. The dermal skeleton consists of loosely connected calcareous bodies (ossicles) with spines or bristles.

In general, fossil finds are rare and usually consist of isolated skeletal plates. The Asterozoa represent the most uniform and durable type of echinoderm. The oldest known remains are from the Ordovician. There are still relatively few species from all the other periods. There are about 3000 Recent species.

1.   Subclass: Somasteroidea

This is an ancient stelleroid group which probably forms the ancestral group of the Asteroidea and Ophiuroidea. The arms are not clearly demarcated from the body disc. There is a double row of semi-cylindrical plates (ambulacrals) and a large pentagonal mouth opening. The water-vascular system runs through a partly or completely closed tube in the ambulacrals. From the ambulacrals parallel rods which consist of individual segments (virgalia) run obliquely to the periphery of the arms.

The Somasteroidea are represented by very few forms from the Lower Ordovician to the present day. The Genus *Villebrunaster* from the Tremadoc (Lowermost Ordovician) of France and the Recent *Platasterias* are examples of the Somasteroidea.

2.   Subclass: Asteroidea (starfishes)

Asteroids have a flat pentactinal body. The wide arms gradually merge into the discoidal body (Fig. 231). The mouth opening lies in the centre of the underside, while the anus is dorsal, i.e. on the upper side, and interradial. The ambulacral system consists of the circum-oral ring and five radial canals. The stone canal, which is sometimes preserved in fossils, runs from the ring canal to the dorsal madreporite.

Movement in asteroids is slow, so they are furnished with a number of protective devices such as spines, thorns, paxillae or pincer-like pedicellariae, all developed as external skeletal processes. The paxillae are calcareous spines on the plates and their free ends are covered with smaller calcareous spines. Respiration is effected by small evaginations of the skin, the so-called papulae (papillae). Asteroids generally have five arms, although there are occasionally forms with up to 40 arms.

Fig. 231. *Calliderma smithiae*, oral surface. Upper Cretaceous, England (×0.25).

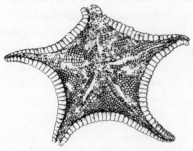

The innumerable tube-feet (with suction cups) which are used for locomotion emanate from the shallow ambulacral grooves on the underside of the animal. The grooves are bounded by the ambulacral plates and may be closed. The ambulacral plates end with an unpaired terminal plate at the tip of the arm. Next to the ambulacral plates lie the adambulacrals, and adjacent to these, the lateral marginals which are usually present in two parallel rows. The lower of these rows consists of the inferomarginals, the upper row of the superomarginals (Fig. 232).

The mouth is surrounded by an oral support consisting of modified ambulacral and interambulacral plates. The main components of the oral support are the mouth-angle plates and the circum-oral plates behind which there is an unpaired odontophore plate. On the upper (aboral) surface there is often a central or dorsocentral plate. A row of large radials extends along the arms, the first and also largest of which is known as the primary radial. There may be a radiocentral plate between this and the central plate. Large interradials are interspersed between the primary radials.

Fig. 232. Subclass Asteroidea: section through the arm of a starfish. Ab, abactinal plate; Ac, actinal plate; Ad, adambulacral; Am, ambulacral; IM, lower marginal plate (inferomarginal); SM, upper marginal plate (superomarginal); T, terminal plate.

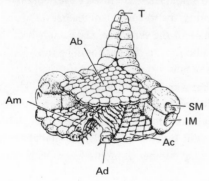

The asteroids are the only echinoderms capable of tolerating temporarily dry conditions, e.g. in the mud-flats of the North Sea. They also have the capacity to regenerate their arms. Asteroid morphology has hardly changed since their appearance in the Ordovician. Fossil finds have mostly consisted of isolated plates which are of stratigraphic value in the Upper Cretaceous of North-West Europe, for example. Reasonably complete specimens from the chalk of North-West Germany have only been found for a few genera: e.g. *Metopaster* (Fig. 233, 2), Cenomanian–Miocene; *Recurvaster* (Fig. 233, 1), Santonian–Danian;

Fig. 233. Subclass Asteroidea: (1) *Recurvaster* (× 0.5); (2) *Metopaster* (× 0.5).

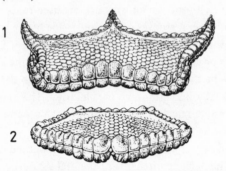

Chomataster, Turonian–Maastrichtian. *Asterias, Oreaster* and *Astropecten* are Recent genera which are quite widespread and common in relatively shallow coastal waters. Some species are found at depths below 3000 m. Similar conditions can be assumed to apply to fossil asteroids.

3.   Subclass: Ophiuroidea (brittle-stars)
        The Ophiuroidea have thin cylindrical arms which are clearly demarcated from the central disc. The arms are very flexible and may be up to 60 cm long and forked in some species. The arms are covered by four rows of plates dorsally, ventrally and laterally. The paired ambulacrals are fused to vertebra-like articulated internal skeletal elements in the arms, the so-called ambulacral vertebrae. The ambulacral canal runs along the base of the vertebrae, and each half of the individual vertebrae sends out a tube-foot to the outside. The tube-feet have no ampullae; they act only as tactile organs.
        The central disc contains the blind-ending gut; there is no anus. The star-shaped, pentactinal mouth is in the centre of the oral side. At each of the corners of the mouth there is a largish plate, the buccal scutum. These oral plates (buccal scuta) are each bounded on the inside by two narrow lateral oral plates (adoral scuta), and adjacent to these an equal number of mouth-angle plates (oral

Fig. 234. Subclass Ophiuroidea: *Ophiocoma*, aboral surface with some parts removed. F, tube-foot; L, lateral plate; O, oral (buccal) plate; S, spine, V, ventral plate; Ve, vertebra (× 3.5).

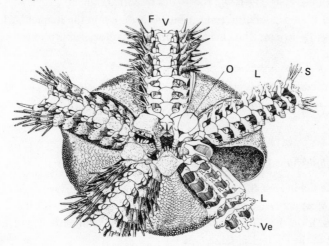

scutella). One of the buccal scuta is perforated and represents the madreporite which allows the entry of water into the stone canal and thus into the water-vascular system (Fig. 234). Brittle-stars have great regenerative powers and are able to discard and regenerate arms and even parts of the central disc.

Ophiuroids have been in existence since the Ordovician. Like the asteroids, they can be derived from the somasteroids.

Depending on the differentiation of the ambulacral plates, four orders are distinguished:

1.   *Order: Stenurida*
         No ambulacral vertebrae. Few genera. Occurrence: Ordovician–Devonian.

2.   *Order: Ophiurida*
         Ambulacral vertebrae present. Diverse order with genera such as *Taeniaster* (syn. *Bundenbachia*) from the Bundenbach Shale (Lower Devonian; Rhineland), which is one of the most common and well preserved fossil ophiuroids, and *Geocoma* from the Solnhofen limestone (Malm ζ).

Ophiocoma, and *Gorgonocephalus* with its extensively branched arms, are examples of Recent genera. Recent vagile benthic ophiurids are found in all seas from shallow waters down to depths approaching 7000 m. They live on sandy and muddy substrates, generally staying in the same area throughout their lives. They usually exhibit a negative phototropic response.

3.   Order: Oegophiurida
The ambulacrals are not fused into vertebrae. The central disc is covered with plates arranged like roof-slates or with a skin.

Furcaster is an oegophiurid genus which is common in the Bundenbach Shale (Lower Devonian; Rhineland). Occurrence: Ordovician–Recent.

4.   Order: Phrynophiurida
The central disc is covered with skin. Large peristomial plates. Many Recent genera such as *Ophiomyxa* are part of this order. Occurrence: Devonian–Recent.

5.   **Subphylum: Echinozoa** (sea-urchins and sea-cucumbers)

The Echinozoa embrace cylindrical to spherical forms with radial symmetry. Most are free-living. They were formerly combined with the Asterozoa to form the 'Eleutherozoa' (Greek: 'the free animals'). This term is now used in the ecological sense only.

According to the *Treatise on Invertebrate Paleontology* (Moore & Teichert, 1952ff) the Echinozoa comprise the following classes:

1.   Class: Helicoplacoidea
Helicoplacus (Fig. 245, 1): This genus, with its spirally arranged and loosely connected plates which apparently formed an extensible test, is the only one in the class. There is a single partly branched spiral ambulacrum. Occurrence: Lower Cambrian.

2.   Class: Ophiocystoidea
Small group with few species, similar to brittle-stars but without arms. Members of this group had very long ventral podia (tube-feet). Occurrence: Lower Ordovician–Middle Devonian.

3.   Class: Cyclocystoidea
Few small flat discoidal forms reminiscent of flat sea-urchins; with long ventral podia. Occurrence: Middle Ordovician–Middle Devonian.

4.   Class: Edrioasteroidea
Known only from Europe and North America. More or less sac-like forms; plates loosely connected. The mouth is in the centre of the upper surface. There are usually five curved ambulacral grooves, e.g. *Edrioaster* (Ordovician: Fig. 235). Members of this group probably led a sessile benthic life (frequently

Fig. 235. Class Edrioasteroidea: *Edrioaster*, oral surface. A, anus (×1).

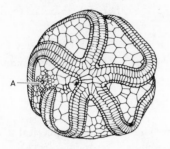

attached to other organisms), although some forms were probably mobile to a limited extent. Occurrence: Lower Cambrian–Lower Carboniferous.

5.   Class: Holothuroidea (sea-cucumbers)

With their elongated bodies, the holothurians diverge considerably from the other echinoderms. The mouth at the front of the animal is surrounded by variably branched tentacles; the anus is terminal. The calcareous skeleton is reduced to small sclerites of very variable shape. These calcareous bodies (10–100 μm long), which may be shaped like hooks, anchors, rings and plates, are assigned to various form genera (Fig. 236).

The animals have gill-like respiratory organs (respiratory trees). Only in a few species is the water-vascular system still in contact with the external environment

Fig. 236. Class Holothuroidea: the Recent holothurian *Chiridota* sp., and the sclerites (*c*. ×150) of various Recent holothurians.

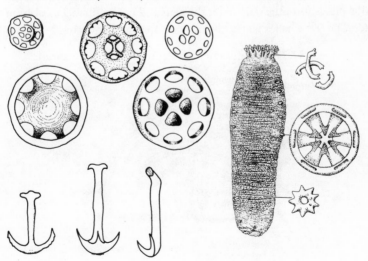

by way of the stone canal and hydropore. Of the five radial canals, three are ventral (trivium) and form the slug-like sole; the other two are dorsal (bivium). The tube-feet are well developed.

Holothurians have probably existed since the Cambrian. Present-day forms are known from both tidal and abyssal zones; in tropical areas they are common in the coastal regions.

6.   Class: Echinoidea (sea-urchins)

Sea-urchins are spiny echinoderms which are almost spherical to discoidal in shape. The test (corona) consists of firmly linked calcite plates. The oral area (peristome) lies centrally or peripherally on the underside; the anal region (periproct) lies on the apex or on the rear half between the apex and the mouth.

According to the position of the mouth and anus two groups are distinguished: regular (Regularia) and irregular (Irregularia) echinoids. In regular echinoids the mouth and anus are opposite each other (Fig. 237); in irregular ones the anus migrates to the back, often as far as the oral side. The mouth may migrate to the front, in which case the test is elongated and bilaterally symmetrical in shape. Since irregular sea-urchins have evolved at least twice from regular ones, the terms regular and irregular are now used only in the descriptive sense and no longer in the systematic sense.

As a rule the apical area consists of ten plates; in regular echinoids it surrounds the anus (endocyclic echinoids), whereas in irregular (exocyclic) ones the anus lies outside the apical region in the rear interambulacral area. Of the ten plates in the apical region the five larger ones, the genital plates, form the upper boundary to the interambulacra (Fig. 238). Sea-urchins are dioecious. The pores in the genital plates form the outlets for the gonads. Fertilisation takes place in the water. The five smaller plates, the ocular plates (also known as terminal plates,

Fig. 237. Organisation of a regular sea-urchin (schematic).

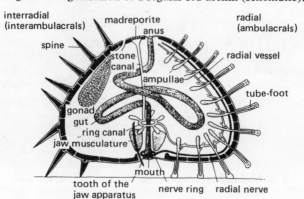

interradial (interambulacrals)   madreporite   radial (ambulacrals)
anus
spine   stone canal   radial vessel
ampullae
tube-foot
gonad
gut
ring canal
jaw musculature
mouth
tooth of the jaw apparatus   nerve ring   radial nerve

Fig. 238. Apical region of a dicyclic regular sea-urchin (spines omitted).
A, anus, G, genital plate; M, madreporic plate; O, ocular plate; PA, pores
of the ambulacral plates.

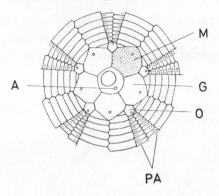

since there is no connection with 'eyes'), are directly on top of the ambulacra
and partly on top of the interambulacra. This type of apical system is known as
dicyclic (Fig. 238). In the monocyclic apical system, the ocular plates push in
between the genital plates and form one ring of plates with them.

The front right genital plate is perforated; it serves as the madreporite and is
in communication with the stone canal which leads to the ring canal round the
mouth. From the mouth five radial canals run upwards under the ambulacra.
From the radial canal tube-feet (organs of locomotion or respiration) emerge
through each pair of pores. The ambulacral and interambulacral areas each
have two alternating rows of plates (Fig. 239). If the pore pairs of the ambulacra

Fig. 239. Test of a regular sea-urchin. A, ambulacrum; IA, interambulacrum.

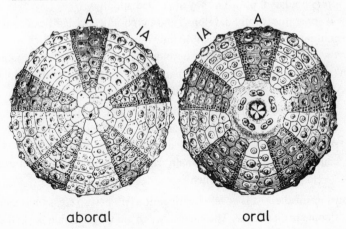

aboral   oral

are joined externally by grooves they are said to be conjugate. While interambulacral plates are always simple, ambulacral plates may be made up of several
individual plates. Various compound ambulacral plates are distinguished,
depending on the pattern: e.g. echinoid, diademoid and cidaroid (Fig. 240).

The shape of the ambulacra is either simple, i.e. a band running from the apex
to the mouth, or leaf-like, i.e. petaloid. In the latter case they form leaf-like areas

Fig. 240. The structure of the ambulacral plates of regular sea-urchins.

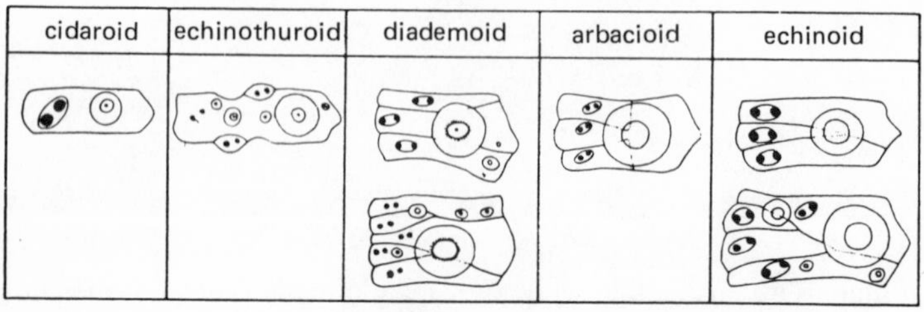

cidaroid	echinothuroid	diademoid	arbacioid	echinoid

('petals') round the apex, or they are open at the bottom and very elongated
in which case they are subpetaloid ambulacra (Fig. 250, 2). Those parts of the
petals surrounded by rows of pores are known as the interstitial areas.

The peristome is covered by a skin-like membrane. It may be holostomatous
(entire) or glyphostomatous (with notches for the mouth gills). The peristome
may be surrounded by a floscelle, a five-rayed star of depressed ambulacra
(known as phyllodes in some irregular echinoids). If a solid masticatory apparatus
('Aristotle's lantern') is present, the inner edge of the peristome is furnished with
five ear-shaped curved processes for the attachment of the 'masticatory muscles'
(Fig. 241). The inner edge of the peristome is known as the perignathic girdle.
If the processes have an ambulacral position, they are known as auricles; if they
have an interambulacral position they are called apophyses.

The jaw apparatus is fully developed and taxonomically significant only in
regular echinoids. It then consists of 40 individual elements of which five are
long narrow teeth. Depending on the development of the teeth, the foramen
magnum (a median notch in the upper part of the pyramid) and the epiphysis,
various types of lantern are distinguished:

cidaroid: teeth longitudinally grooved, foramen shallow;
aulodont: teeth longitudinally grooved, foramen fairly deep;
stirodont: teeth longitudinally keeled, foramen fairly deep;
camarodont: teeth longitudinally keeled, foramen deep, epiphysis
    closed above the foramen

The surface of the test is covered with various processes, e.g. spines and
the microscopic pedicellariae. The spines, which are very variable in their shape

Fig. 241. Jaw apparatus of a regular sea-urchin ('Aristotle's lantern').

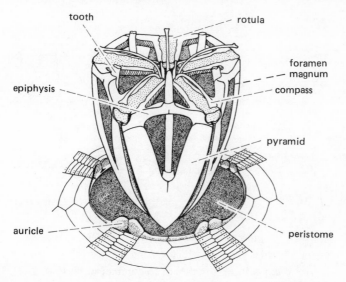

and histology, are attached by ball and socket joints and moved by the action of muscles. They serve the purpose of protection, locomotion and the catching of prey. The ball-joints (tubercles, mamelons) may be pitted on top for the attachment of a ligament and surrounded by a ring of smaller tubercles. The base of the tubercle is surrounded by a smooth area, the areola (Fig. 242).

The pedicellariae are stalked and bear tripartite pincers, which may be modified into snapping pincers (tridactylar pedicellariae), serrated biting pincers (trifoliate pedicellariae) and poisonous pincers (gemmiform pedicellariae). In Recent forms they are used as diagnostic criteria, whereas fossil pedicellariae are identifiable only in exceptional cases.

Fig. 242. *Plegiocidaris*: primary spines and interambulacral plates with primary tubercles (× 0.8).

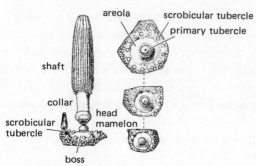

Fig. 243. *Spatangus*: (a) plastron (×0.7); (b) various fascioles (×1).

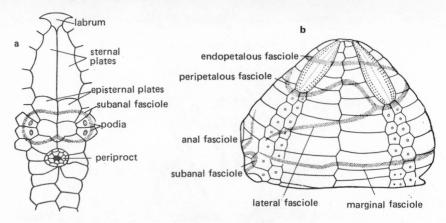

Fascioles, which are present only in the holasteroids and spatangoids, are areas covered with bristles and cilia which extend over certain parts of the surface and serve to expel foreign bodies or to conduct water (Fig. 243). The enlarged rear interambulacrum on the oral side of holasteroids and spatangoids is known as the plastron.

Orientation in the echinoids follows the system of Lovén (1874). The plane of bilateral symmetry (Lovén's plane) runs through the front radius (III) and the rear interradius (5). The bilateral symmetry is more obvious in the irregular echinoids where the anus migrates to (5). The madreporite lies in sector (2) (Fig. 244).

Ecology

All 800 Recent species are marine and benthic. They inhabit all parts of the ocean from the coast to the abyssal zones. The spatangoids live down to depths exceeding 7000 m. The greatest diversity in echinoids is to be found in littoral tropical and subtropical zones.

Sea-urchins graze freely on hard or soft substrates on the sea-bottom, as well as in cracks and caves, digging in the muddy or sandy substrate or occasionally boring into rocks. The spatangoids, for example, burrow as deep as 20 cm with their spines, whilst keeping a canal open with specially modified feet so that respiratory water is able to reach them. A waste canal often ends blind in the substrate (e.g. in *Echinocardium*). In the burrowing spatangoids the fascioles create currents of water towards the mouth or away from the anus. Irregular echinoids mostly live buried in the sand, and in favourable localities the population density may be high.

Fig. 244. I: Orientation in the echinoderms according to Carpenter's system (for pelmatozoans, but applicable to all echinoderms). The oral surface is turned towards the observer. H, hydropore or madreporic plate. II: The orientation of holothurians according to Carpenter's system. III: The orientation of a sea-urchin. The commonly used orientation (after Lovén) is indicated by numerals, the corresponding terms according to Carpenter's system are indicated by letters. The oral surface is directed towards the observer. M, madreporic plate (projected position, as it is actually on the aboral surface). Stippled areas: ambulacra.

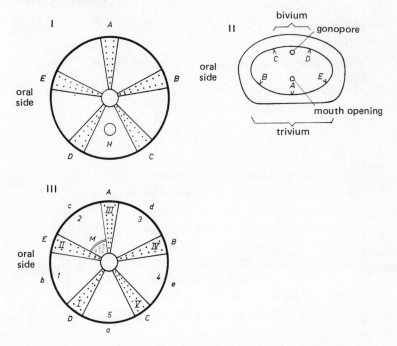

Sea-urchins may be omnivorous, carnivorous or herbivorous (feeding on algae, sessile invertebrates and dead animals). While some forms, e.g. members of the Order Clypeasteroida (so-called sand-dollars), feed microphagously by sifting the sand with their spines and removing the organic constituents, other forms pick up organic particles from the sediment with specialised tube-feet.

A brief phylogenetic history

Sea-urchins have been in existence since the Ordovician and may be derived from the edrioasteroids. Characteristics developed after the Ordovician period are:

1. the development of a rigid test;

2. the development of a preferred direction of locomotion and the corresponding orientation;
3. the improvement of the water-vascular system;
4. the modification of the tube-feet for special functions;
5. specialised feeding habits.

The Palaeozoic echinoids were small. At the end of the Palaeozoic all but the line of the cidaroids became extinct. In the Upper Triassic the first 'modern' regular sea-urchins appeared, and their number of species subsequently increased dramatically. The first irregular sea-urchins appeared in the Lower Jurassic.

Overview of classification

1. Subclass: Perischoechinoidea
   Ambulacral areas consisting of two to 20 columns of plates, interambulacral areas consisting of one to many columns.
   Order: Bothriocidaroida (Ordovician)
   Order: Echinocystitoida (Ordovician–Permian)
   Order: Palaechinoida (Silurian–Permian)
   Order: Cidaroida (Upper Silurian–Recent)
2. Subclass: Euechinoidea
   Occurrence: Triassic–Recent. Ambulacral and interambulacral areas consisting of two columns of plates.
   Superorder: Diadematacea (Triassic–Recent) ⎱ 'Regularia'
   Superorder: Echinacea (Triassic–Recent) ⎰
   Superorder: Gnathostomata (Jurassic–Recent) ⎱ 'Irregularia'
   Superorder: Atelostomata (Jurassic–Recent) ⎰

Examples:

1. Subclass: Perischoechinoidea
    Regular echinoids with variable number of ambulacral and interambulacral columns.

1.   Order: Bothriocidaroida
      Bothriocidaris (Fig. 245, 2): Ambulacra consisting of two columns, interambulacra of one column of plates. Madreporite radial. Rigid test. No genital plates. Up to 1 cm in diameter.

2.   Order: Echinocystitoida
      Ambulacra consisting of 2–20 columns of plates, the interambulacra of 1–14 columns; plates overlapping like roof-slates; flexible test. *Aulechinus*

Fig. 245. Subphylum Echinozoa, Class Helicoplacoidea: (1) *Helico-placus* (×1.4). Class Echinoidea, Subclass Perischoechinoidea: (2) *Bothriocidaris* (×1.5); (3) *Melonechinus* (×1.5); (4) *Eothuria* (×1.5); (5) *Aulechinus* (×1.5).

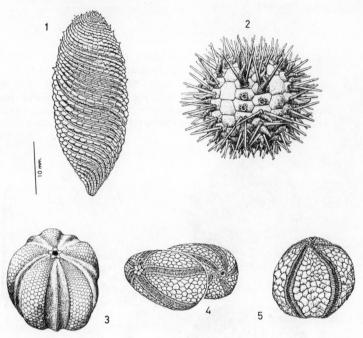

and *Eothuria* (Fig. 245, 4, 5) are two well known genera from the Ordovician of England.

3.   Order: Palaechinoida

Ambulacra consisting of two or more columns of plates, interambulacra of one or more columns. Corona rigid and spherical with rib-like raised areas. *Melonechinus* (Fig. 245, 3) is known from the Lower Carboniferous of Europe and North America.

4.   Order: Cidaroida

Test initially consisting of overlapping plates; rigid test since the Triassic; usually 20 columns of plates, 30 columns in older forms. No auricles, only apophyses present. One single large tubercle on each interambulacral plate. The only Recent order to have survived from the Palaeozoic. The order went through its prime in the Mesozoic. These days, members of this order are particularly common in the Indo-Pacific area and in the Antarctic.

Archaeocidaris (Fig. 246, 1): Interambulacra consisting of four columns
of plates (= pluriserial); plates slightly mobile; with prominent perforated
primary tubercles surrounded by a granulated ring. Cylindrical spines. Occurrence:
Carboniferous, Permian.

Fig. 246. Subclass Perischoechinoidea: (1) *Archaeocidaris* (× 0.7);
(2) *Tylocidaris* spines (× 0.7); (4) *Rhabdocidaris*, with spine (a) (× 0.5).
Subclass Euechinoidea: (3) *Phymosoma* (× 0.5), with spine (a, enlarged);
(5) *Salenia*, (a) seen from the side, (b) magnified apical disc, (c) view
from above (× 1); (6) *Stomechinus* (× 0.5); (7) *Hemicidaris* (× 0.7),
with apical system; (8) *Glypticus* (× 1.5).

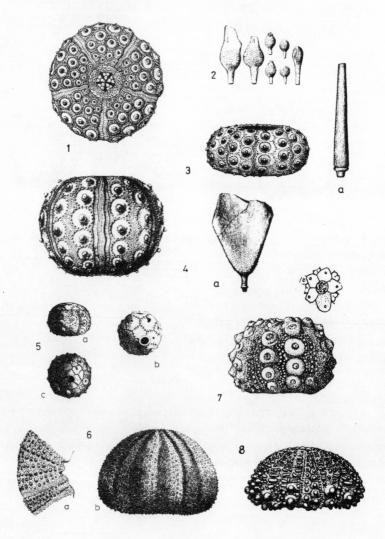

Cidaris: Test slightly flattened at the top and bottom. Holostomatous oral area; curved ambulacra; pores not conjugate; areolae usually deep. Small and numerous ambulacral plates. Occurrence: Recent.

Rhabdocidaris (Fig. 246, 4): Large test (up to 10 cm in diameter) slightly flattened at the top and bottom. Ambulacral areas clearly curved. Conjugate pores; long thorny spines. Occurrence: Jurassic (Lias)-Eocene.

Tylocidaris (Fig. 246, 2): Primary tubercles not perforated; pores not conjugate. Very large club-shaped spines, which are sometimes of biostrati-graphic importance in the Upper Cretaceous. Occurrence: Cretaceous-Danian.

2.   Subclass: Euechinoidea
Rigid test consisting of five double columns each of ambulacral and interambulacral plates. Anus and periproct either endocyclic (within the apical disc) or secondarily displaced to interambulacral sector (5) (exocyclic). Occurrence: Upper Triassic-Recent.

The regular members of the two following superorders are also combined with the Perischoechinoidea as the 'Regularia'.

1.   *Superorder: Diadematacea*
Test almost radially symmetrical. Plates firmly linked or overlapping like roof-slates. Occurrence: Triassic-Recent.

1.   *Order: Echinothurioida*
Aulodont jaw apparatus (cf. Cidaroida). Glyphostomatous oral area; diademoid ambulacral plates. The periproct lies within the apical disc; five genital pores present. Many present-day abyssal forms. Occurrence: Triassic-Recent.

Echinothuria (Fig. 247, 2): A representative with a large test exceeding 10 cm in diameter and consisting of thin plates. Occurrence: Upper Cretaceous-Recent.

2.   *Order: Diadematoida*
The plates of the corona are firmly linked or overlap like roof-slates. The primary tubercles are usually notched; the primary spine is hollow along its axis. The periproct lies within the apical disc. Occurrence: Triassic-Recent.

Eodiadema: Flattened test. Simple ambulacral areas; monocyclic apical disc. Long spines. Occurrence: Jurassic (Lias).

Fig. 247. Subclass Euechinoidea: (1) *Pygaster*, (a) aboral, (b) oral, (c) lateral (×0.5); (2) *Echinothuria*, aboral ambulacral plates (×0.6).

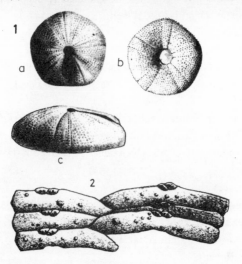

3.   Order: Pygasteroida

Medium-sized; with rigid corona. The apical system touches the periproct, but genital plate 5 is absent, and genital plate 3 and ocular plates II–IV are outside the apical disc. Pits present in the adoral interambulacral plates. Spines with a solid axis. Occurrence: Jurassic–Cretaceous.

Pygaster (Fig. 247, 1): Test rounded or pentagonal. Apical side convex, oral side flat with sunken peristome. Well developed madreporite; third genital plate no longer in contact with the periproct. Occurrence: Jurassic (Bathonian)–Cretaceous (Cenomanian).

Pseudodiadema: Diademoid ambulacral plates (Fig. 240). Large, clearly glyphostomatous peristome. Striated spines. Occurrence: Upper Triassic (Rhaetian)–Lower Cretaceous (Aptian).

2.   Superorder: Echinacea

Plates firmly linked, not imbricate. Periproct within the apical disc. Perignathic girdle and jaw apparatus present. Occurrence: Triassic–Recent.

1.   Order: Salenioida

Endocyclic periproct; jaw apparatus present; compound ambulacral plates of the echinoid type. Occurrence: Lower Jurassic–Recent.

Salenia (Fig. 246, 5): Peristome rounded; apex with a dorsocentral (suranal) plate; anal opening eccentric. Rounded peristome with weak notches, surrounded by ten perforated buccal plates. Primary tubercles imperforate. Possibly a persistent juvenile form. Occurrence: Cretaceous-Recent.

2. *Order: Hemicidaroida*
    Hemicidaris (Fig. 246, 7): On the underside the ambulacra have two rows of primary tubercles which become smaller on the upper surface. No suranal plate present. Ambulacra slightly curved and much narrower than the interambulacra. Very large spines. Occurrence: Jurassic-Cretaceous.

3. *Order: Phymosomatoida*
    Phymosoma (Fig. 246, 3): Complex ambulacral plates with numerous pores; pores biserial on the apical side. Primary tubercles imperforate, notched. Occurrence: Jurassic (Malm)-Tertiary (Eocene).
    This order also includes *Stomechinus* (Fig. 246, 6). Occurrence: Lower Jurassic-Lower Cretaceous.

4. *Order: Arbacioida*
    Glypticus (Fig. 246, 8): Narrow ambulacra with two rows of tubercles; interambulacra with irregular 'disrupted' tubercles. Apical disc dicyclic. Primary tubercles imperforate, not notched. Occurrence: Upper Jurassic.

5. *Order: Echinoida*
    Psammechinus: Regularly hemispherical test. Each ambulacral plate with three obliquely arranged pores. Secondary tubercles very numerous. Apical disc regular. Occurrence: ? Miocene-Recent.

The two following superorders may also be combined as the 'Irregularia'.

3. *Superorder: Gnathostomata*
    Exocyclic periproct, but usually still situated on the aboral surface. Hollow spines. Primary tubercles generally perforated and notched. Jaw apparatus present. Occurrence: Lower Jurassic-Recent.

1. *Order: Holectypoida*
    Compound ambulacral plates of the diademoid type; generally small; probably of monophyletic origin. Occurrence: Jurassic-Tertiary.

Holectypus (Fig. 248, 1): Small, low, rounded test. Simple ambulacral area. Large periproct on the more or less flat oral side. The primary tubercles

Fig. 248. Subclass Euechinoidea: (1) *Holectypus* (x 0.35); (2) *Galerites* (x 0.7); (3) *Clypeaster* (x 0.4); (4) *Scutella* (x 0.4). A, anus, M, mouth.

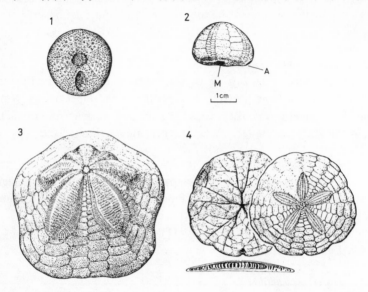

are perforated and notched and larger and more numerous on the oral surface than on the aboral surface. Four genital plates. The peristome has marked slits. Occurrence: Jurassic–Cretaceous.

Conoclypus: Large thick-walled test with an oval outline. Pronounced petals on the aboral side. Wide interambulacral areas. The periproct is inframarginal and very small. Small tubercles and spines. Occurrence: Tertiary.

Galerites (syn. *Echinoconus*) (Fig. 248, 2): Medium-sized, conical to hemispherical test with a flattened oral surface. The central peristome has a rounded outline. Primary tubercles perforated and notched. The periproct lies below the ambitus. Occurrence: Upper Cretaceous.

2.   *Order: Clypeasteroida*
       The test is generally flattened dorsoventrally; inside there is a system of well developed internal supports. Petaloid ambulacral areas on the aboral surface. The position of the periproct is very variable. Lunules may be developed as keyhole-shaped perforations from the oral through the aboral surface of the test. Other perforations or indentations may be present, as in the case of extremely flat forms (sand-dollars). Occurrence: Cretaceous (Maastrichtian)– Recent.

Clypeaster (Fig. 248, 3): Wide pronounced petals in which ambulacral plates alternate with demiplates. Ambulacral areas sometimes fused at the peristome. Well developed internal supports inside the test. Madreporite fused with the genital plates; five genital pores. The periproct is marginal or infra-marginal. Occurrence: Tertiary-Recent. Of stratigraphical significance in the Upper Tertiary.

Scutella (Fig. 248, 4): Strongly flattened discoidal test up to 15 cm in diameter. No indentations or perforations (lunules). Small peristome surrounded by a rosette of ten wedge-shaped plates. Small inframarginal periproct. Apical disc with four genital pores. Occurrence: Tertiary.

4. *Superorder: Atelostomata*
Irregular echinoids with no jaw apparatus or perignathic girdle. The peristome is holostomatous, the periproct is exocyclic. Primary spines hollow. Occurrence: Jurassic (Lias)-Recent.

1. *Order: Cassiduloida*
Irregular echinoids of very variable shape with petals on the apical side and phyllodes and floscelles on the oral side. The jaw apparatus is resorbed in the course of ontogeny. Most forms are known from the Eocene and Miocene. Occurrence: Jurassic-Recent.

Echinolampas (Fig. 250, 2): Oval test with petals which are open at the ends. More or less prominent floscelle; four genital pores. Oblique-oval inframarginal periproct. Simple spines. Occurrence: Tertiary (Eocene)-Recent.

Clypeus (Fig. 250, 1): Flattened test, circular in outline. Petaloid ambulacra, open at the bottom and with conjugate pores. The pentagonal peristome is eccentric and surrounded by a floscelle. Large madreporite and four small genital plates. Occurrence: Dogger (Bajocian)-Malm (Kimmeridgian).

2. *Order: Holasteroida*
Elongated or disrupted apical disc. Floscelle absent. Fascioles variable. Occurrence: Jurassic (Lias)-Recent.

Collyrites (Fig. 250, 3): Oval inflated test. Simple ambulacra. Oval and marginal periproct. The front four perforated genital plates and three ocular plates are separated from the rear two ocular plates by two intercalated radial plates on the apex. This gives rise to a front trivium and a rear bivium. Occurrence: Dogger–Malm.

Holaster (Fig. 250, 6): Test very convex, oval heart-shaped in outline. Front ambulacrum in a shallow groove. Pores not conjugate. Elongated apical disc; second and fourth ocular plates fused, thus separating the front genital plates from the rear ones; four genital pores. Peristome elliptical, periproct oval. Occurrence: Cretaceous–Tertiary.

Echinocorys (Fig. 250, 7): Test very convex, egg-shaped in outline. Ambulacra with tiny, slightly conjugate pores. Test covered with small tubercles. Elliptical peristome with two lips, situated in the anterior portion of the flat oral surface. The oval periproct is close to the rear margin. Occurrence: Upper Cretaceous–Danian.

Infulaster (Fig. 249, 1): Conical test, front end with a prominent groove accommodating ambulacrum III. The periproct is at the tip of the rear indentation; peristome anterior. Marginal fasciole. Occurrence: Turonian.

Hagenowia (Fig. 249, 2): Similar to *Infulaster* but with a slender rostrum. Ambulacrum III deeply sunken. Occurrence: Coniacian–Maastrichtian.

Offaster (Fig. 249, 3): Similar, almost spherical test; ambulacra not sunken or petaloid. Periproct above the marginal fasciole. Occurrence: Upper Cretaceous.

3.   *Order: Spatangoida*
Elongated, heart-shaped tests with sunken petals. Fascioles present in most forms (Fig. 243). Occurrence: Lower Cretaceous–Recent. Climax of evolution in the Tertiary (Eocene).

Micraster (Fig. 249, 5): Heart-shaped outline. A shallow groove running from the apex to the anterior margin. Paired petaloid ambulacra. Conjugate pores. Apical disc with four genital pores. Peristome displaced towards the front, with prominent lip (labrum). The periproct is situated on the truncated posterior end; below there is a subanal fasciole. Occurrence: Upper Cretaceous–Tertiary.

Spatangus (Fig. 249, 4): Test with heart-shaped outline; subanal fasciole characteristically present. The plastron (flattened area formed from the modified posterior interambulacrum) is completely covered with tubercles. Very large madreporic plate. Lives buried in the sand. Occurrence: Tertiary (Eocene)– Recent.

Fig. 249. Subclass Euechinoidea: (1) *Infulaster* (×0.5). (a) lateral,
(b) oral, (c) frontal; (2) *Hagenowia* (×0.6); (3) *Offaster*, (a) aboral,
(b) oral, (c) lateral (×0.5); (4) *Spatangus*, (a) aboral (×0.4), (b) apical
disc (enlarged: G, genital plate; M, madreporic plate; O, ocular plate);
(5) *Micraster* (×0.5), with apical system; (6) *Echinocardium*, (a) view
from behind, (b) aboral (a, b: ×0.7), (c) in life position in the sediment
(×0.25).

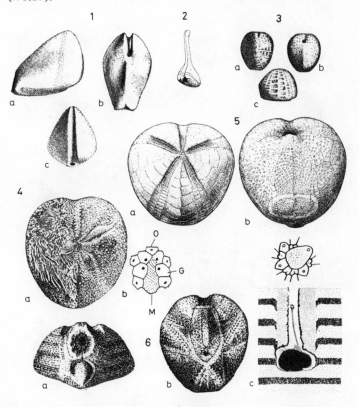

Echinocardium (Fig. 249, 6): Heart-shaped, thin-walled test. The
anterior ambulacrum and the apical disc are surrounded by an endopetalous
fasciole. Apical side covered with small tubercles. Lives buried in the sand of
the littoral zone. Occurrence: Tertiary (Oligocene)–Recent.

Toxaster (Fig. 250, 4): Heart-shaped test. Wide groove running from
the apical disc to the anterior edge, bounded by conjugate pores. Ambulacra
curved and open with elongated pores. Four genital pores on the apical disc.
Pentagonal peristome without labrum; oval periproct. Small tubercles. Fascioles
absent. Occurrence: Cretaceous.

Fig. 250. Subclass Euechinoidea: (1) *Clypeus* (×0.25); (2) *Echinolampas* (×0.7); (3) *Collyrites* (×0.5); (4) *Toxaster* (×0.7); (5) *Maretia* (×0.5); (6) *Holaster* (×0.5); (7) *Echinocorys* (×0.4), with apical system.

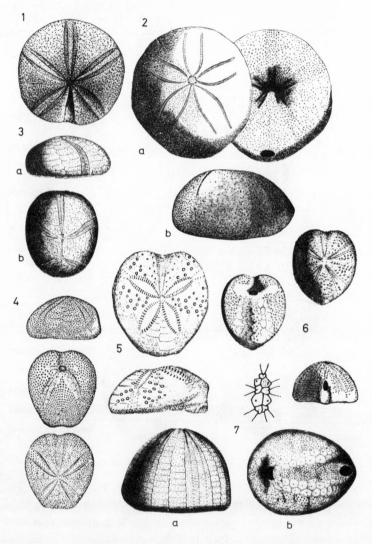

Hemiaster: Short, domed test with a shallow groove running from the apical disc to the anterior edge. The groove is surrounded by round, paired, conjugate pores. Petaloid ambulacra, of which the four posterior ones are much shorter than the anterior ones. All are surrounded by a peripetalous fasciole. The apical disc is central with an asymmetrical madreporite. Forms of this genus show a marked tendency towards losing their bilateral symmetry. Occurrence: Cretaceous (Aptian)-Recent.

Maretia (Fig. 250, 5); Test with subanal fasciole. Anterior ambulacrum not petaloid. Plastron covered with tubercles. Occurrence: Tertiary-Recent.

18. Phylum: Branchiotremata (Stomochordata, Hemichordata)

Small deuterostomes with internal branchial baskets.

Since this phylum has rather inconspicuous members and is represented by only a few species, it plays a rather minor role in the present-day fauna. Even so, it is of great phylogenetic significance for two reasons: firstly, it is either the ancestor or an early branch of the vertebrates; and secondly, it includes the extinct graptolites which provide by far the most important and 'accurate' guide fossils for the Ordovician and Silurian.

1. Class: Enteropneusta (acorn worms)

Acorn worms are several centimetres to over 2 m in length. They live in passages they have dug themselves and cemented with slime. The front portion of the digestive canal is modified into a gill-like organ, a network of narrow slits which exit dorsally and are supported by skeletal rods of connective tissue. This organ serves the purposes of respiration and gathering food. Branchial baskets of this description are only known in the chordates. The tube-shaped inturned front portion of the fore-gut, which is reminiscent of a notochord, also suggests a relationship with the chordates.

About 70 species of acorn worm are known. None has preservable hard parts, so there are no fossil representatives.

The genus *Balanoglossus* includes the largest species *B. gigas* whose representatives have a diameter of only 2 cm in spite of being up to 2.5 m long.

2. Class: Pterobranchia

The Pterobranchia are small animals measuring only 2-3 mm, in exceptional cases up to 14 mm. With one exception, all forms live in epibenthic colonies in solid tubes secreted by the animals themselves. The individuals within a colony are connected by stolons. The branchial basket which is well developed in the Enteropneusta is reduced to two slits in the Pterobranchia. On the dorsal side there are several arms, each furnished with a double row of ciliated tentacles which collect food and are probably also used for respiration.

The most diverse genus of the Pterobranchia, *Cephalodiscus*, is restricted to the southern hemisphere (16 species), while the genus *Rhabdopleura* (Fig. 251) predominantly inhabits the northern hemisphere (3 species).

Fig. 251. Organisation of a *Rhabdopleura* colony (schematic).

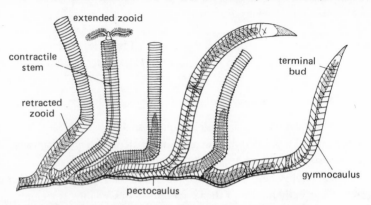

The tubes are built up from a secretion supplied by glands on the head. The secretion is spread over the end of the tube in the form of a half-ring, so that the tubes are made up of half-rings. These are joined along zigzag sutures. Contrary to earlier reports, the building material is not chitin but a scleroprotein.

A few fossil Pterobranchia have been described from the Ordovician of Poland.

3.   Class: Graptolithina (graptolites)
The graptolites constructed tube-like skeletons similar to those of the Pterobranchia, which also consisted of half-rings joined together by zigzag sutures. With increasing age these were covered with an ever thicker periderm layer. The building material was a scleroprotein consisting of the same amino acids as that of the Pterobranchia but of a slightly different composition.

The pointed conical sicula is the first-formed chamber of the colony (rhabdosome). All other thecae originate from the sicula by budding, although the sicula itself is of sexual origin. The entire colony is penetrated by a thread-like soft body (stolon) from which the individuals (zooids) branch off in tube-shaped or cup-shaped thecae.

The sicula consists of the larval chamber (prosicula) and the metasicula, which resembles the thecae to be formed later. In the live animal the opening of the sicula was directed downwards. The sicula terminates in a long apical thread-like nema (also known as virgula) from which hangs the rhabdosome. The sicula is 1–3 mm long (Fig. 252).

The most primitive graptolites were sessile, either branching in a tree-like fashion or of irregular shape (Fig. 253). The rhabdosome consisted of three types of theca which occurred in regular patterns: the largish autotheca (presumably the female zooid), the smaller bitheca (for the male zooid) and

Fig. 252. Structure of the sicula in *Monograptus*.

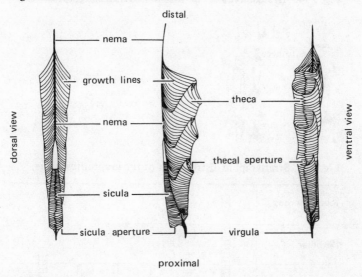

the stolotheca which contained and extended the stolon but did not actually appear as a theca on the fully developed rhabdosome (Fig. 254). With the transition to a planktonic lifestyle the bithecae were reduced, so that only the autothecae remained visible. Presumably their inhabitants were hermaphrodites. Occurrence: Cambrian–Carboniferous (Fig. 255).

Fig. 253. Reconstruction of a dendroid graptolite.

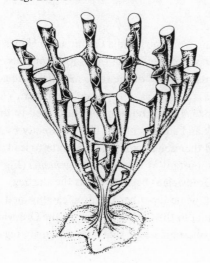

Fig. 254. Arrangement of the thecae in dendroid graptolites.

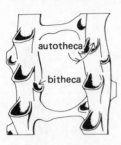

Fig. 255. Stratigraphic distribution of the graptolite orders.

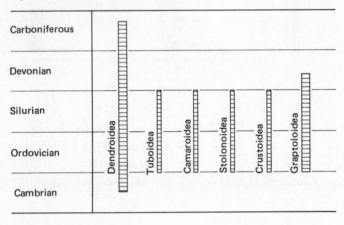

1.   Order: Dendroidea

The oldest order, occurring from the Middle Cambrian to Upper
Carboniferous. The rhabdosome was branched like a tree with numerous
branches connected by dissepiments made up of the same material as the
periderm. The three types of theca - autotheca, bitheca and stolotheca -
formed simple tubes of different size. The rhabdosome was attached to the
substrate by means of a short stalk and an attachment disc or directly by the
sicula. Only a few genera detached themselves from the substrate to lead
a hemiplanktonic life. The most important of these was *Dictyonema* (Fig. 256)
from the Tremadoc (Lowermost Ordovician) because it was the starting
point for subsequent development, while the other Dendroidea remained
unchanged until they became extinct in the Carboniferous. In the Ordovician
and Silurian there were also a few aberrant sessile side lines which are regarded
as orders in their own right:

Fig. 256. Order Dendroidea: *Dictyonema* (×1).

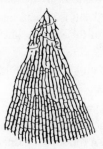

2.   *Order: Tuboidea*
         With irregular budding.

3.   *Order: Camaroidea*
         With basally inflated autothecae.

4.   *Order: Stolonoidea*
         With hypertrophied stolons.

5.   *Order: Crustoidea*
         Encrusting graptolites; the autothecae have specially differentiated
apertures while the stolothecae retain a simple tube shape. Occurrence: Ordovician.
         Within the family of the Dendrograptidae, which is represented by *Dictyonema*
and which spread all over the world, rapid evolution ensued even within the
Tremadocian Age which led to a reduction of the number of thecae and stipes.
These simplified Dendroidea gave rise to the Order Graptoloidea.

6.   *Order: Graptoloidea*
         Members of this order were exclusively planktonic or hemiplanktonic.
Only the most primitive forms still had bithecae in addition to the autothecae.
Thanks to their solid casing the stolons of the Dendroidea could be preserved,
whereas graptoloid stolons are unknown because they had no preservable casing.
The reduction in the number of stipes was associated with increased differentiation of thecal architecture which affected the peristomes in particular.
         The building material of the graptolite rhabdosome was elastic but able to
retain its shape. Following the reduction in the number of stipes from more
than four to two, various forms appeared with pendent or horizontal stipes
(with respect to the sicula) or with stipes raised to varying degrees (reclined,
scandent). In some cases the stipes actually fused over the sicula so that forms
arose with biserial thecae. They are characteristic of the Upper Ordovician.

At the Ordovician–Silurian boundary some forms reduced the thecae on one side, beginning at the proximal end (proterogenetically), so that uniserial forms arose. The latter make up the graptolite fauna of the Silurian and Lower Devonian.

The graptoloids are divided into two not always easily distinguishable groups based on the configuration of the stipes:

1. Both stipes remain free, so that the rhabdosome is bilateral and uniserial. The nema which emanates from the sicula also remains free. This group is collectively known as the Axonolipa, a rather outdated but commonly used term.
2. The stipes are erect (scandent) and fused back to back with the nema (also known as virgula in this configuration) in between. The rhabdosome is thus biserial or secondarily uniserial. These forms are known as Axonophora.

Of the following families the first three are Axonolipa while the others are Axonophora:

1.   *Family: Dichograptidae*
With two to many stipes; their angle of divergence is usually less than 180°; with simple tube-like thecae.

Dichograptus (Fig. 257, 1): Rhabdosome branched dichotomously, with a maximum of eight stipes which are long and flexible; thecae usually strongly overlapping. Occurrence: Lower Ordovician.

Tetragraptus (Fig. 257, 2): Four stipes; thecae simple, notched. Probably a polyphyletic genus. Occurrence: Lower Ordovician.

Didymograptus (Fig. 257, 3): Only two stipes; thecae usually simple, straight, at most with a slight ventral curve. Occurrence: Lower to Middle Ordovician.

2.   *Family: Leptograptidae*
The angle of divergence is 180°; the stipes bear thecae shaped like a pipe-bowl (leptograptid type).

Leptograptus (Fig. 258, 1): Two stipes, diverging at an angle of 180°; long thecae of the leptograptid type. Occurrence: Middle to Upper Ordovician.

Fig. 257. Order Graptoloidea: (1) *Dichograptus* (× 0.35); (2) various species of *Tetragraptus* (a–c: × 1; d, e: × 4); (3) two species of *Didymograptus* (a, b: × 0.7).

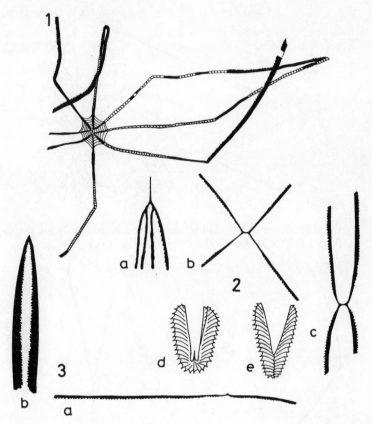

Nemagraptus (Fig. 258, 2): Two, usually S-shaped main stipes which may give rise to branches from their convex side in a regular pattern. Thecae of the leptograptid type. Occurrence: Middle Ordovician.

3.   *Family: Dicranograptidae*

The two branches are scandent and initially united and biserial; they then diverge and become uniserial. The apertures of the thecae are curved inwards.

Dicellograptus (Fig. 259, 1): Thecae of the leptograptid type. Occurrence: Middle to Upper Ordovician.

Fig. 258. Order Graptoloidea: (1) various species of *Leptograptus* (a, b: ×0.5; c: ×1.5; d: ×4); (2) *Nemagraptus* (×0.5).

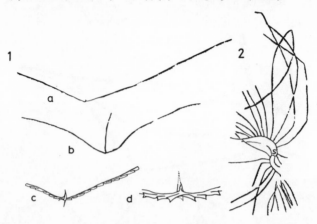

Fig. 259. Order Graptoloidea: various species of (1) *Dicellograptus* (a: ×0.5; b: ×2) and (2) *Dicranograptus* (a: ×0.5; b: ×2).

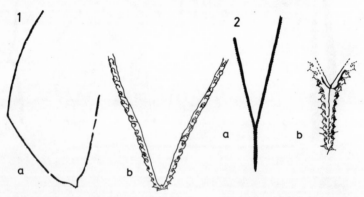

Dicranograptus (Fig. 259, 2): Thecae strongly incurved. Occurrence: Middle Ordovician.

4.   *Family: Diplograptidae*

With scandent stipes which have fused and thus given rise to a biserial rhabdosome. The thecae are simple or of the leptograptid type. A number of genera are distinguished on the basis of the rhabdosome cross-section.

Climacograptus (Fig. 260, 1): Thecal apertures in small indentations of the rhabdosome; rhabdosome cross-section almost circular. Occurrence: Lower Ordovician–Lower Silurian.

Fig. 260. Order Graptoloidea: various species of (1) *Climacograptus* (a, c, d: × 1; b: × 3) and (2) *Diplograptus* (a, b: × 1).

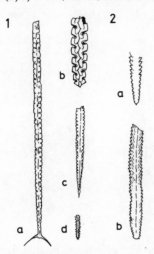

Diplograptus (Fig. 260, 2): Thecae straight to sigmoidal. Rhabdosome cross-section egg-shaped to rectangular. Occurrence: Middle Ordovician–Lower Silurian.

5.　*Family: Cryptograptidae*

Rhabdosome biserial because of the lateral contact between two stipes; often covered with spines.

Glossograptus (Fig. 261, 1): With long lateral and apertural spines. Long straight thecae. Occurrence: Lower to Middle Ordovician.

Cryptograptus (Fig. 261, 2): Occurrence: Lower to Upper Ordovician.

Fig. 261. Order Graptoloidea: (1) *Glossograptus hincksi* (different views; × 1); (2) *Cryptograptus* (× 1).

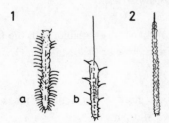

Fig. 262. Order Graptoloidea: (1) *Retiolites*, (a) rhabdosome (×1),
(b) structure (×6); (2) two species of *Dimorphograptus* (×1).

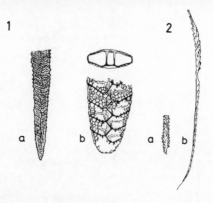

6.   *Family: Retiolitidae*
    The periderm (exoskeleton) is reduced to a network. Biserial rhabdosome.
Polyphyletic family.
    The genus *Retiolites* (Fig. 262, 1) occurred from the Lower to Upper Silurian.

7.   *Family: Dimorphograptidae*
    The rhabdosome is uniserial at the proximal (sicula) end and biserial at
the distal end as a result of the unilateral reduction of the thecae from the
proximal end.

    Dimorphograptus (Fig. 262, 2): Thecae straight or slightly incurved.
Occurrence: Lower Silurian.

8.   *Family: Monograptidae*
    The rhabdosome is completely uniserial. In this family the thecae are
extensively modified.

1.   *Subfamily: Monograptinae*
    Simple rhabdosome. *Monograptus, Rastrites.*

    Monograptus (Fig. 263, 1): Thecal shape very variable, with the thecae
at the proximal end sometimes differing from those at the distal end. Occurrence:
Silurian-Lower Devonian.

    Rastrites (Fig. 263, 2). Rhabdosome more or less incurved; thecae very
long and widely spaced. Occurrence: Lower Silurian.

Fig. 263. Order Graptoloidea: (1) *Monograptus* species from the
Lower Silurian, (a) *M. cyphus* (×1), (b) *M. bohemicus* (×1),
(c) *M. dubius* (×1), (d) *M. priodon* (×1), (e) *M. convolutus* (×1),
(f) *M. discus* (×2). (2) Different species of *Rastrites* (×1).

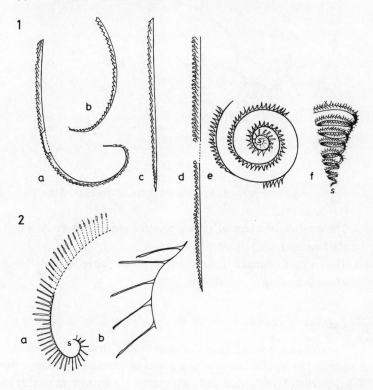

2. **Subfamily: Cyrtograptinae**
Rhabdosome branched. *Cyrtograptus.*

Cyrtograptus (Fig. 264): Rhabdosome coiled, with lateral branches
(cladia) arising from thecal apertures. Occurrence: Middle Silurian.

Evolutionary lines

According to H. Jaeger (1978) the evolution of the graptolites can be
divided into four main steps:

1. The transition from the sessile to the planktonic lifestyle at the
   beginning of the Ordovician.
2. The transition from the dimorphic thecae of the Dendroidea to the

Fig. 264. Order Graptoloidea: *Cyrtograptus* (× 1).

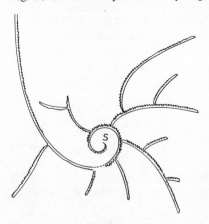

monomorphic thecae of the Graptoloidea at the end of the Tremadoc (Ordovician).

3. The creation of a biserial axonophore rhabdosome in the Arenig (Ordovician).

4. The creation of the uniserial axonophore *Monograptus* rhabdosome at the beginning of the Silurian.

Ecology

Graptolites are particularly abundant in dark shales ('black carbonaceous shales'), but are also found in limestones or even in sandstones, often very well preserved. Many genera and even species are distributed worldwide, while others display a certain degree of provincialism. Since they could not have survived in the euxinic conditions at the bottom of the black shale seas

Fig. 265. Synrhabdosome of *Glossograptus*, (a) from above, (b) from the side (× 0.5).

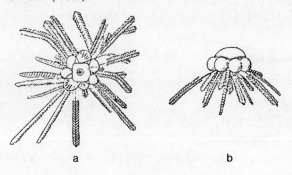

a   b

they must have lived in the upper layers of water. These days they are regarded as planktonic, since this is the only explanation for their wide distribution. Some forms have been found with several rhabdosomes aggregated (synrhabdosome) and kept afloat by a communal flotation apparatus (Fig. 265). The presence of other flotation devices such as long spines, reduction of the wall thickness, wider nemae, special coiling, etc., also suggests a planktonic floating lifestyle. The relationship of the graptolites with the Pterobranchia finally suggests that the graptolite zooids may also have had tentacles with cirri whose action would have lifted the whole rhabdosome, as well as serving the purpose of collecting food.

The graptolites have proved to be particularly valuable guide fossils thanks to their wide regional distribution, their very rapid evolution in some cases, their large number of individuals and the relatively easy identification even of badly preserved rhabdosomes.

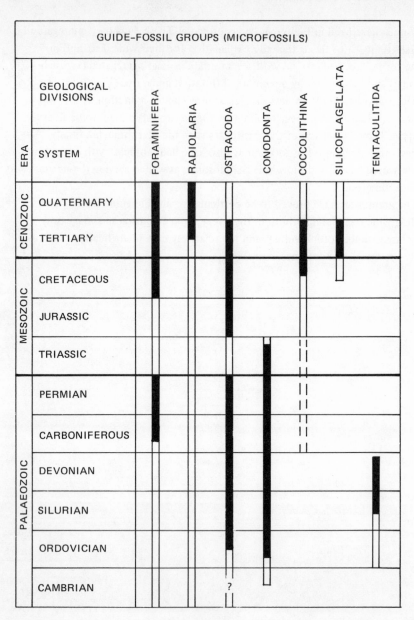

GUIDE-FOSSIL GROUPS (MICROFOSSILS)

GUIDE-FOSSIL GROUPS (MACROFOSSILS)											
GEOLOGICAL DIVISIONS		TRILOBITA	GRAPTOLITHINA	ANTHOZOA	BRACHIOPODA	NAUTILOIDEA	AMMONITOIDEA	BELEMNOIDEA	LAMELLIBRANCHIA	GASTROPODA	MAMMALIA	VASCULAR PLANTS
ERA	SYSTEM											
CENOZOIC	QUATERNARY											
	TERTIARY											
MESOZOIC	CRETACEOUS											
	JURASSIC											
	TRIASSIC											
PALAEOZOIC	PERMIAN											
	CARBONIFEROUS											
	DEVONIAN											
	SILURIAN											
	ORDOVICIAN											
	CAMBRIAN											

EON	ERA (ERATHEM)	SYSTEM (PERIOD) ('FORMATION')	EPOCH, STAGE				MILLION YEARS
PHANEROZOIC	CENOZOIC	QUATERNARY			HOLOCENE		0.01
					PLEISTOCENE		1.8
		TERTIARY	NEO-GENE		PLIOCENE		
					MIOCENE		
					OLIGOCENE		
			PALAEO-GENE		EOCENE		
					PALEOCENE (DANIAN)		65
	MESOZOIC	CRETACEOUS	UPPER		MAASTRICHTIAN		135
					CAMPANIAN		
					SANTONIAN		
					CONIACIAN		
					TURONIAN		
					CENOMANIAN		
			LOWER		ALBIAN		
					APTIAN		
					BARREMIAN		
					HAUTERIVIAN	NEO-COMIAN	
					VALANGINIAN		
					BERRIASIAN		
		JURASSIC	UPPER	MALM	TITHONIAN		
					KIMMERIDGIAN		
					OXFORDIAN		
			MIDDLE	DOGGER	CALLOVIAN		
					BATHONIAN		
					BAJOCIAN		
					AALENIAN		
			LOWER	LIAS	TOARCIAN		
					PLIENSBACHIAN		
					SINEMURIAN		
					HETTANGIAN		190
		TRIASSIC	UPPER	RHAETIAN	KEUPER		
				NORIAN			
				CARNIAN			
			MIDDLE	LADINIAN	MUSCHEL-KALK		
				ANISIAN			
			LOWER	SCYTHIAN	BUNT-SANDSTEIN		225

EON	ERA	SYSTEM			EPOCH, STAGE			MILLION YEARS
PHANEROZOIC	PALAEOZOIC	PERMIAN		UPPER	ZECHSTEIN		TATARIAN	
							KAZANIAN	
				LOWER	ROT-LIEGENDES		KUNGURIAN	
							ARTINSKIAN	
							SAKMARIAN	280
		CARBONI-FEROUS	PENN-SYLVANIAN	UPPER	STEPHANIAN	URALIAN		
					WESTPHALIAN	MOSCOVIAN & BASHKIRIAN		
					NAMURIAN			
			MISSIS-SIPPIAN	LOWER	VISEAN	DINANTIAN		
					TOURNAISIAN			345
		DEVONIAN		UPPER	FAMENNIAN			
					FRASNIAN			
				MIDDLE	GIVETIAN			
					EIFELIAN			
				LOWER	EMSIAN			
					SIEGENIAN			
					GEDINNIAN			395
		SILURIAN		UPPER	LUDLOW			
				MIDDLE	WENLOCK			
				LOWER	LLANDOVERY			430
		ORDOVICIAN		UPPER	ASHGILL			
					CARADOC			
					LLANDEILO			
					LLANVIRN			
				LOWER	ARENIG			
					TREMADOC			500
		CAMBRIAN			UPPER			
					MIDDLE			
					LOWER			570
CRYPTOZOIC	PRECAMBRIAN	EOCAMBRIAN		VENDIAN RIPHEAN	ALGONKIAN			
		= PROTEROZOIC						2000
		ARCHAEAN						4500 – 5000

Bibliography

The bibliography includes only a selection of relevant titles which may be consulted for further lists of references.

Ager, D. V. (1963) *Principles of Paleoecology.* McGraw-Hill, New York.

Beklemischew, W. N. (1958–60) *Grundlagen der vergleichenden Anatomie der Wirbellosen,* 2 vols. VEB Deutscher Verlag der Wissenschaften, Berlin.

Bengtson, St. (1976) The structure of some middle Cambrian conodonts and the early evolution of conodont structure and function. *Lethaia,* **9**, 185–206.

Black, R. (1970) *The Elements of Palaeontology.* Cambridge University Press, Cambridge.

Boardman, R. S., Cheetham, A. H. & Oliver, W. A. (1973) *Animal Colonies: Development and Function through Time.* Dowden, Hutchinson & Ross, Stroudsburg, Pennsylvania.

Brasier, M. D. (1980) *Microfossils.* Allen & Unwin, London.

British Museum (1969) *British Palaeozoic Fossils,* 3rd edn. British Museum (Natural History), London.

   (1975a) *British Mesozoic Fossils,* 5th edn. British Museum (Natural History), London.

   (1975b) *British Caenozoic Fossils,* 5th edn. British Museum (Natural History), London.

Callomon, J. H. (1963) Sexual dimorphism in Jurassic ammonites. *Trans. Leicester Lit. Phil. Soc.,* **57**, 36 pp.

Evitt, W. R. (1963) A discussion and proposals concerning fossil dinoflagellates, hystrichospheres and acritarchs. *Proc. Natn. Acad. Sci.,* **49**, 158–64, 298–302.

Flügel, E. (1978) *Mikrofazielle Untersuchungsmethoden von Kalken.* Springer Verlag, Berlin.

Gothan, W. & Weyland, H. (1973) *Lehrbuch der Paläobotanik,* 3rd edn. BLU, Munich, Berne, Vienna.

Grassé, P. P. (ed.) (1948ff) *Traité de zoologie.* Masson, Paris (to be completed).

Hahn, G. & Pflug, H. D. (1980) Ein neuer Medusenfund aus dem Jung-Präkambrium von Zentral-Iran. *Senckenbergiana Lethaia,* **60**(4/6), 449–61.

Haq, B. U. & Boersma, A. (eds.) (1978) *Introduction to Marine Micropaleontology.* Elsevier, Amsterdam.

Hecker, R. F. (1965) *Introduction to Paleoecology.* Elsevier, New York.

Imbrie, J. & Newell, N. D. (eds.) (1964) *Approaches to Paleoecology.* Wiley, New York.

Jaeger, H. (1978) Entwicklungszüge (Trends) in der Evolution der Graptolithen. *Schriftenr. geol. Wiss.,* 5–58.

Jefferies, R. P. S. (1979) Calcichordates. In *The Encyclopedia of Paleontology,* ed. R. W. Fairbridge & D. Jablonski, pp. 161–7. Dowden, Hutchinson & Ross, Stroudsburg, Pennsylvania.

Jenkins, R. J. F., Plummer, P. S. & Moriarty, K. C. (1981) Late Precambrian pseudofossils from the Flinders Ranges, South Australia. *Trans. R. Soc. S. Australia,* **105**(2), 67–83.

Krumbiegel, G. & Walther, H. (1977) *Fossilien.* F. Enke, Stuttgart.

Kunth, A. (1869/78) Beiträge zur Kenntnis fossiler Korallen. *Z. Dtsch. geol. Ges.,* **21**, 647–87; **22**, 24–43.

Lauterbach, K.-E. (1980) Schlüsselereignisse in der Evolution des Grundplans der Arachnata (Arthropoda). *Abh. naturw. Ver. Hamburg* (NF), **23**, 163–327.

Lehmann, U. (1976) *Ammoniten – ihr Leben und ihre Umwelt.* F. Enke, Stuttgart. (Published in English by Cambridge University Press as *The Ammonites* (1981).)

Lehmann, U. (1977) *Paläontologisches Wörterbuch,* 2nd edn. F. Enke, Stuttgart.

Lovén, S. L. (1874) *Etudes sur les Echinoidées,* Vol. II. Svenska Vetensk. Handl., Stockholm.

Lowenstam, H. A. (1963) Biologic problems relating to the composition and diagenesis of sediments. In *The Earth Sciences: Problems and Progress in Current Research,* ed. T. W. Donley, pp. 137–95. University of Chicago Press.

Makowski, H. (1963) Problem of sexual dimorphism in ammonites. *Palaeontol. Polonica,* **12**, 1–92.

Milliman, J. D. (1974) *Marine Carbonates: Recent Sedimentary Carbonates,* pt 1. Springer, Berlin.

Moore, R. C. & Teichert, C. (eds.) (1952ff) *Treatise on Invertebrate Paleontology.* McGraw-Hill, New York.

Moore, R. C. & Teichert, C. (eds.) (1952ff) *Treatise on Invertebrate Paleontology.* Geological Society of America and University of Kansas Press, Lawrence, Kansas (to be completed):
Part A *Introduction* (1979)
Part C *Protista 2* (1964)
Part D *Protista 3* (1954)
Part E *Archaeocyatha and Porifera:* 1st edn (1955); 2nd edn (revised and enlarged), vol. 1 *Archaeocyatha* (1972)
Part F *Coelenterata* (1956)
Part G *Bryozoa* (1953)
Part H *Brachiopoda:* vol. 1 (1956); vol. 2 (1965)
Part I *Mollusca 1* (1960)
Part K *Mollusca 3* (1964)
Part L *Mollusca 4* (1957)
Part N *Mollusca 6:* vol. 1 *Bivalvia* (1969); vol. 2 *Bivalvia* (1969); vol. 3 *Bivalvia* (1971)
Part O *Arthropoda 1* (1959)
Part P *Arthropoda 2* (1955)
Part Q *Arthropoda 3* (1961)
Part R *Arthropoda 4:* vols 1 and 2 (1969)
Part S *Echinodermata 1:* vols 1 and 2 (1967 [1968])
Part T *Echinodermata 2: Crinoidea,* vols 1–3 (1978)
Part U *Echinodermata 3:* vols 1 and 2 (1966)
Part V *Graptolithina:* 1st edn (1955); 2nd edn (revised and enlarged, 1970)
Part W *Miscellanea* (1962); Supplement 1 (1955)

Müller, A. H. (1965–1980) *Lehrbuch der Paläozoologie.* Vol. I, *Allgemeine Grundlagen* (1976); Vol. II, *Invertebraten,* pt 1 (1980), pt 2 (1965), pt 3 (1978). VEB G. Fischer, Jena.

Pander, C. (1856) Monographie der fossilen Fische des silurischen Systems der russisch-baltischen Gouvernements. *Akad. Wiss. St Petersburg,* **I–X**, 1–91.

Pflug, H. D. (1966) Einige Reste niederer Pflanzen aus dem Algonkium. *Palaeontographica, Abt. B,* **117**(4–6), 59–74.

Pflug, H. D. (1979) Combined structural and chemical analysis of 3800-Myr-old micro-fossils. *Nature,* **280**, 483–6.

Piveteau, J. (ed.) (1952–69) *Traité de paléontologie*. Masson, Paris.
Pojeta, J., Jr, Runnegar, B. & Kříž, J. (1973) *Fordilla troyensis* Barrande: the oldest known pelecypod. *Science*, **180**, 866–8.
Rassmussen Wienberg, H. (1969) *Palaeontologi – Fossile Invertebraten*. Munksgaard, Copenhagen.
Raup, D. M. & Stanley, St. M. (1978) *Principles of Paleontology*, 2nd edn. W. H. Freeman, San Francisco.
Reid, R. E. H. (1950) A monograph of the Upper Cretaceous Hexactinellida of Great Britain and Northern Ireland. *Pal. Soc. Lond.* **111**, pt I, 1–46; pt II, 1–26; pt III, 1–48; pt IV, 1–154.
Remane, A., Storch, V. & Welsch, U. (1976) *Systematische Zoologie: Stämme des Tierreichs*. G. Fischer, Stuttgart.
Ryland, J. S. (1970) *Bryozoans*. Hutchinson, London.
Schäfer, W. (1962) *Aktuo-Paläontologie nach Studien in der Nordsee*. Kramer, Frankfurt.
Schindewolf, O. H. (1962) Studien zur Stammesgeschichte der Ammoniten: II. *Abh. Akad. Wiss. Lit. Mainz, Math.-Naturwiss. Kl.*, **8**, 113–258.
Tasch, P. (1978) *Paleobiology of the Invertebrates: Data Retrieval from the Fossil Record*, 2nd edn. Wiley, New York.
Thenius, E. (1976) *Allgemeine Paläontologie*. Prugg Verlag Wien-Eisenstadt; Verlag Brüder Hollinek, Vienna.
Trauth, F. (1927–36) Aptychenstudien: I–VIII. *Ann. Naturhist. Mus. Wien*, **41–47**.
Trauth, F. (1938) Die Lamellaptychi des Oberjura und der Unterkreide. *Palaeontographica*, **88**, 115–229.
Wedekind, R. (1918) Die Genera der Palaeoammonoidea (Goniatiten). *Palaeontographica*, **A62**, 85–184.
Wenz, W. & Zilch, A. (1959/60) *Gastropoda*, pt 2, *Euthyneura*. Bornträger, Berlin.
Whittaker, R. H. (1969) New concepts of kingdoms of organisms. *Science*, **163**, 150–60.
Wiedmann, J. (1973) Evolution or revolution of ammonoids at mesozoic system boundaries? *Biol. Rev.*, **48**, 159–94.
Wurmbach, H. (1971) *Lehrbuch der Zoologie*, vol. II, *Spezielle Zoologie*. G. Fischer, Stuttgart.
Ziegler, B. (1972) *Allgemeine Paläontologie: Einführung in die Palökologie*, pt 1. Schweizerbart'sche Verlagsbuchhandlung, Stuttgart.
Zittel, K. A. (1915) *Grundzüge der Paläontologie*, section I, *Invertebrata*, 4th edn. Munich & Berlin.

Acknowledgements

The illustrations acknowledged here have been left unchanged in only a few cases; most have been redrawn, simplified, extended or changed in some other way. Illustrations made up of several individual ones may be derived from a number of sources. Those illustrations not listed below are based on the authors' own designs.

Alberti, G. K. G. (1975) Zur Struktur der Gehäusewand von *Styliolina* (Dacryoconarida) aus dem Unter-Devon von Oberfranken. *Senckenbergiana lethaea*, 55. Figs. 143, 144

Black, R. (1970) *The Elements of Palaeontology*. Cambridge University Press, Cambridge. Figs. 50; 52A, 6; 52B, 10, 11a–c; 193; 246; 249; 250

Brasier, M. D. (1980) *Microfossils*. Allen & Unwin, London. Figs. 6; 7; 14; 16B; 20a–c, e; 20d (after Pokorny); 6; 7; 14; 16; 20; 160B; 161; 212a–i

Cowen, R. & Rider, J. (1972) Functional analysis of fenestellid bryozoan colonies. *Lethaia*, 5. Fig. 184

Cuffey, R. J. (1967) Bryozoan *Tabulipora carbonaria* in Wreford Megacyclothem (Lower Permian) of Kansas. *University of Kansas Paleontological Contributions:* Bryozoa. Article 1. Fig. 179

Cushman, J. A. (1978) *Foraminifera, their Classification and Economic Use*. Harvard University Press, Cambridge, Mass. Fig. 19

Erben, H. K. (1975) *Die Entwicklung der Lebewesen. Spielregeln der Evolution*. R. Piper, Munich & Zurich. Fig. 2

Ernst, G. (1964) Ontogenie, Phylogenie und Stratigraphie der Belemnitengattung *Gonioteuthis* Bayle aus dem nordwestdeutschen Santon/Campan. *Fortschr. Geol. Rheinl. Westfalen*, 7. Fig. 132

Ernst, G. (1972) Grundfragen der Stammesgeschichte bei irregulären Echiniden der nordwesteuropäischen Oberkreide. *Geol. Jahrb. Reihe*, 4. Fig. 248

Flor, F. (1970) Demonstrationen an rezenten und fossilen Tintenfischen. *Der Biologie-Unterricht*, Jg. 6, 3. E. Klett, Stuttgart. Fig. 125

Fraas, E. (1910) *Der Petrefaktensammler*. K. G. Lutz, Stuttgart (reprinted 1972). Fig. 83

Fry, W. G. (ed.) (1970) *The Biology of the Porifera*. Academic Press, New York & London. Fig. 33

Gabel, B. (1971) Die Foraminiferen der Nordsee. *Helgol. Wiss. Meeresunters.*, 22, 1–65. Fig. 12

Glaessner, M. F. (1971) Die Entwicklung des Lebens im Präkambrium und seine geologische Bedeutung. *Geol. Rundsch.*, 60(4), 1323–39. Fig. 5

Gothan, W. & Weyland, H. (1973) *Lehrbuch der Paläobotanik*, 3rd edn. BLU, Munich, Berne & Vienna. Figs. 9, 10

Götting, K.-J. (1974) *Malakozoologie*. G. Fischer, Stuttgart. Fig. 58

Gross, W. (1957) Über die Basis der Conodonten. *Paläontol. Z.*, 31. Fig. 210

Handlirsch, A. (1925) *Fossile Insekten*. In *Handbuch der Entomologie*, 5th issue, ed. Ch. Schröder. Figs. 170, 171

Hess, H. (1975) Die fossilen Echinodermen des Schweizer Jura. *Veröffentl. Naturhist. Museum Basel*, 8. Figs. 223, 228, 234, 236, 242, 246

Hillmer, G. (1971) Bryozoen – Tierkolonie im Meer. *Mikrokosmos*, 3, Fig. 175

Jaekel, O. (1918) Phylogenie und System der Pelmatozoen. *Paläontol. Z.,* **3.**
Figs. 222, 230

Janus, H. (1958) *Unsere Schnecken und Muscheln.* Kosmos-Naturführer,
Franckh'sche Verlagsbuchhandlung, Stuttgart. Fig. 70

Kaestner, A. (1965) *Lehrbuch der speziellen Zoologie.* VEB G. Fischer, Jena.
Fig. 62

Kazmierćzak, J. (1979) Sclerosponge nature of chaetetids evidenced by
spiculated *Chaetetopsis favrei* (Deninger 1906) from the Barremian of
Crimea. *N. Jahrb. Geol. Paläontol. Mh.,* **2.** Fig. 34

Krömmelbein, K. (1977) *Historische Geologie.* In *Abriss der Geologie,* vol. 2, ed.
Brinkmann. F. Enke, Stuttgart. Table 3; Figs. 107, 118

Krummbiegel, G. & Walther, H. (1977) *Fossilien.* F. Enke, Stuttgart. Figs. 63,
195

Kühn, A. (1959) *Grundriss der Allgemeinen Zoologie.* G. Thieme, Stuttgart
Figs. 18, 237

Lauterbach, K. S. (1980) Schlüsselereignisse in der Evolution des Grundplans der
Arachnata (Arthropoda). *Abh. naturw. Ver. Hamburg* (NF), **23.** Figure on
p. 190

Lehmann, U. (1976) *Ammoniten -ihr Leben und ihre Umwelt.* Ferdinand Enke,
Stuttgart. (Published in English by Cambridge University Press (1981).
Figs. 91, 101, 119, 120, 122, 123, 126

Lehmann, U. (1977) *Paläontologisches Wörterbuch,* 2nd edn. F. Enke, Stuttgart.
Figs. 1, 28, 47, 53, 92, 129, 160C, 163, 195–197, 241, 244

Lindström, M. (1964) *Conodonts.* Elsevier, Amsterdam. Fig. 211

McKinney, F. (1977) Autozooecial budding patterns in dendroid Paleozoic
bryozoans. *J. Paleontol.,* **51,** 2. Figs. 182, 183

Mauritz, H.-J. (1975) Vorkommen, Morphologie und Entwicklungstendenzen
fossiler eukaryotischer Algen. Unpublished thesis, Hamburg. Fig. 8

Moore, R. C., Lalicker, C. G. & Fischer, A. G. (1952) *Invertebrate Fossils.*
McGraw-Hill, New York. Figs. 93–95, 181, 184–186, 194, 216, 218, 230

Moore, R. C. & Teichert, C. (eds.) (1952ff) *Treatise on Invertebrate Paleontology.*
Geological Society of America and University of Kansas Press, Lawrence,
Kansas (to be completed). Figs. 15, 17, 22, 30, 31, 35, 36, 39, 40, 41,
42, 52, 60, 61, 64, 65, 66, 67, 77, 78, 79, 80, 81, 82, 84, 85, 86, 87, 88,
89, 90, 96, 97, 102, 103, 106, 108, 109, 110, 111, 112, 113, 114, 116,
117, 119, 120, 121, 141, 146, 147, 149, 150, 151, 152, 153, 154, 155,
156, 158, 160, 168, 169, 176, 177, 178, 180, 182, 185, 187, 190, 191,
205, 206, 207, 209, 214, 217, 235, 245, 247, 249, 250, 251, 256–265

Müller, A. H. (1965–1980) *Lehrbuch der Paläozoologie,* Vol. II, *Invertebraten.*
G. Fischer, Jena. Figs. 53, 73, 74, 127, 130–132, 135, 239, 240, 246

Naef, A. (1922) *Die fossilen Tintenfische.* Jena. Fig. 124

Pflug, H. D. (1974) Vor- und Frühgeschichte der Metazoen. *N. Jahrb. Geol.
Paläontol. Abh.,* **145**(3), 328–74. Fig. 1

Piveteau, J. (ed.) (1952) *Traité de paléontologie,* vol. II. Masson, Paris.
Figs. 198, 203, 205, 208, 243

Pokorny, V. (1958) *Grundzüge der zoologischen Mikropaläontologie,* vol. I.
Berlin. Figs. 19, 25a, b

Rabeder, G. & Steininger, F. (1972) *Leitfaden zu einem paläontologischen
Praktikum (Wirbellose) für Anfänger.* Palaeontological Institute of the
University of Vienna. Figs. 46, 136

Rasmussen Wienberg, H. (1969) Palaeontologi – Fossile Invertebraten. Munksgaard, Copenhagen. Figs. 41, 42, 45, 52, 54, 56, 67, 68, 69, 76, 78, 79,

81, 83, 87, 88, 89, 96a and b, 97, 98, 103, 116, 137, 138, 152, 157,
160–162, 174, 188, 189, 198–202, 206, 217, 223, 225, 227, 229, 230,
231, 233, 245, 246, 248, 249

Remane, A., Storch, V. & Welsch, U. (1976) *Systematische Zoologie: Stämme des Tierreichs.* G. Fischer, Stuttgart.   Figs. 27, 37, 43, 44, 57, 59, 71

Rickards, R. B. & Palmer, D. C. (1977) *Graptolites I.* SA Baldwin, Educational Palaeontological Reproductions, London.   Figs. 252–254

Ryland, J. S. (1970) *Bryozoans.* Hutchinson University Library, London. Fig. 188

Schäfer, W. (1962) *Aktuo-Paläontologie nach Studien in der Nordsee.* W. Kramer, Frankfurt.   Fig. 249

Schidlowski, M. (1971) Probleme der atmosphärischen Evolution im Präkambrium. *Geol. Rundsch.,* **60**(4), 1351–84.   Fig. 3

Schindewolf, O. H. (1950) *Grundfragen der Paläontologie.* E. Schweizerbart'sche Verlagsbuchhandlung, Stuttgart.   Figs. 50A, 55

Schindewolf, O. H. (1955) Zur Taxionomie und Nomenklatur der Clymenien. Ein Epilog. *N. Jahrb. Geol. Paläontol. Mh.*   Fig. 105

Schmidt, M. (1928) *Die Lebewelt unserer Trias.* Hohenlohesche Buchhandlung Öhringen.   Figs. 96b, 111, 207

Schulz, H. D. (1967) *Beloceras* mit ceratitischen Loben aus der Montagne Noire, Frankreich. *N. Jahrb. Geol. Paläontol. Mh.,* **10**, 608–19.   Fig. 103

Schumacher, H. (1976) *Korallenriffe, ihre Verbreitung, Tierwelt und Ökologie.* BLV Verlagsgesellschaft, Munich, Berne & Vienna.   Fig. 49

Seibold, E. (1974) *Der Meeresboden. Ergebnisse und Probleme der Meeres-Biologie.* Springer, Berlin, Heidelberg & New York.   Figs. 24, 48

Spaeth, Chr. (1975) Zur Frage der Schwimmverhältnisse bei Belemniten in Abhängigkeit vom Primärgefüge der Hartteile. *Paläontol. Z.,* **49**, 3. Fig. 128

Stromer v. Reichenbach, E. (1909) *Lehrbuch der Paläozoologie,* pt I, *Wirbellose Tiere.*   Figs. 89, 250

Thenius, E. (1977) *Meere und Länder im Wechsel der Zeiten.* Springer, Berlin, Heidelberg & New York.   Fig. 148

Wanner, J. (1924) *Die permischen Echinodermen von Timor,* pt II, *Paläontologie von Timor.* Issue 14, Abh. 23. Stuttgart.   Fig. 215

Wenz, W. (1938–44) *Gastropoda.* Verlag von Gebrüder Borntraeger, Berlin. Figs. 66, 10; 68, 2, 3, 11; 69

Werner, B. (1967) *Stephanoscyphus* Allman (Scyphozoa, Coronatae), ein rezenter Vertreter der Conulata? *Paläontol. Z.,* **41**, 137–53.   Fig. 38

Whittaker, R. H. (1969) New concepts of kingdoms of organisms. *Science,* **163**, 150–60.   Fig. 4

Wiedmann, J. (1969) The heteromorphs and ammonoid extinction. *Biol. Rev.,* **44.**   Figs. 99, 115

Wurmbach, H. (1971) *Lehrbuch der Zoologie,* vol. II, *Spezielle Zoologie.* G. Fischer, Stuttgart.   Fig. 13

Ziegelmeier, E. (1957) Die Muscheln (Bivalvia) der deutschen Meeresgebiete. *Helgol. Wiss. Meeresunters.,* **6**, 1. List, Sylt.   Fig. 72

Ziegler, B. (1975) *Einführung in die Paläobiologie,* pt 1, *Allgemeine Paläontologie.* E. Schweizerbart'sche Verlagsbuchhandlung, Stuttgart.   Figs. 51, 213

Zittel, K. A. (1915) *Grundzüge der Paläontologie,* sect. I, *Invertebrata,* 4th edn. Munich & Berlin.   Figs. 21, 29, 31, 32, 34, 54, 56, 86, 90, 96a and b, 98, 110, 164–167, 199, 200, 204, 224, 226, 229, 246, 247, 250

Index

Page numbers in *italic* type refer to illustrations

nematocyst 55, 58
Nematomorpha 182
nematopore 223
Neoammonoidea 146, 156, 161
neobelemnites 179
Neocephalopoda 135, 143–79
Neocrioceras 159
Neogastropoda (neogastropods) *86*, 89, 95, 100-2
Neohibolites 179
Neoloricata 84
Neomegalodon 130
Neopilina 85, 85, 86
Neoschwagerina 25, 26
neotrematous *239*, 244
nephridium 86
Nephriticeras 141, 141
Nerinea 90, 97, 98
nerineids 98
Nerinella 90
Nerita 94, 95, 96
neritaceans 96
Neuropora 51
Nevadia 196
Nielsenicrinus 274
Nodosaria 19, 27, 28
Nodosinella 19
Notaspidea 102
Nothocoeli see Duvaliina
notochord 305
notothyrium 243
Nowakia 181
Nubecularia 19
Nucula 109, 110, 112, 113
Nuculana 113
Nuculanidae 113
Nuculida see Palaeotaxodonta
Nuculidae (nuculids) 113
Nudibranchia 102, 103
nummulites 29, 30, 31
Nummulites 19, 30, 31
nummulitic limestone 30
nymph 109

Obolus 221, 238, 244, 245
occipital ring 192, *192*
ocelli 204
Octobrachia see Octopoda
Octocorallia (octocorals) 9, 61, 62-3
Octopoda (octopods) 134, 173
Octopus 173
octopuses see Octopoda
ocular plate 288, *289*
ocular ridge 193
Odonata 219
Odontobelus 177, 177
odontophore 283

Odontopleura 203, 203
Odontopleurida *203*, 203
oeciopore *225*
Oegophiurida (oegophiurids) 286
Offaster 302, 303
Ogygopsis 198, 198
Oldhamina 221, 238, 247, 248, 256
Olenellida (olenellids) 190, *190*, 191, 198
Olenellus 197, 197
Olenidae 200
Olenus 199, 200
Omphalocyclus 29, 31, 32
Oncoceras 140, 140
Oncocerida 140
Onousoecia 225
Onverwacht series 1, *2,* 13
onychite 174
Onychocella 221
Onychophora 190
ooecium see ovicell
operculum 88, 223, *223*
opesium 233
Ophiamyxa 286
Ophioceras 141, 141
Ophiocoma 285, 285
Ophiocystoidea 286
Ophiurida 285
Ophiuroidea (ophiuroids) *264*, 282, 284-6
Opisthobranchia *101*, 102, 103
opisthodetic ligament 109
opisthogyral umbo 109
opisthoparian facial suture 193, *193*
opisthosoma 204, *204, 205,* 206
opisthothorax 190
Oppelia 162, 164
oppeliids 169
oral scutellum 284
Orbiculoida 238, 244, 245
Orbitoides 31, 31, 32
orbitoids 29, 30-2
Orbitolina 19, 23, 23
Orbitolinidae 23
Orbitolites 26, 29
Orbulina 27, 28
Oreaster 284
organ-pipe coral see *Tubipora*
Orthacea 245, 246
Orthida (orthids) *237, 238,* 242, 245-6, 248
Orthis 221, 238, 245, 245, 246
Orthoceras 137, 138, 139, 140
orthoceratids 173
Orthocerida 139, 143
orthochoanitic 139, *139*
orthocone 136
Orthonychia see *Platyceras*
orthotriaene *45*